Practical SCADA
for Industry

Titles in the series

Practical Cleanrooms: Technologies and Facilities (David Conway)

Practical Data Acquisition for Instrumentation and Control Systems (John Park, Steve Mackay)

Practical Data Communications for Instrumentation and Control (John Park, Steve Mackay, Edwin Wright)

Practical Digital Signal Processing for Engineers and Technicians (Edmund Lai)

Practical Electrical Network Automation and Communication Systems (Cobus Strauss)

Practical Embedded Controllers (John Park)

Practical Fiber Optics (David Bailey, Edwin Wright)

Practical Industrial Data Networks: Design, Installation and Troubleshooting (Steve Mackay, Edwin Wright, John Park, Deon Reynders)

Practical Industrial Safety, Risk Assessment and Shutdown Systems (Dave Macdonald)

Practical Modern SCADA Protocols: DNP3, 60870.5 and Related Systems (Gordon Clarke, Deon Reynders)

Practical Radio Engineering and Telemetry for Industry (David Bailey)

Practical SCADA for Industry (David Bailey, Edwin Wright)

Practical TCP/IP and Ethernet Networking (Deon Reynders, Edwin Wright)

Practical Variable Speed Drives and Power Electronics (Malcolm Barnes)

Practical SCADA
for Industry

David Bailey BEng, Bailey and Associates, Perth, Australia

Edwin Wright MIPENZ, BSc(Hons), BSc(Elec Eng), IDC Technologies, Perth, Australia

AMSTERDAM • BOSTON • HEIDELBERG • LONDON • NEW YORK • OXFORD
PARIS • SAN DIEGO • SAN FRANCISCO • SINGAPORE • SYDNEY • TOKYO
Newnes is an imprint of Elsevier

Newnes

Newnes is an imprint of Elsevier
Linacre House, Jordan Hill, Oxford OX2 8DP, UK
30 Corporate Drive, Suite 400, Burlington, MA 01803, USA

First published 2003
Reprinted 2005, 2006

British Library Cataloguing in Publication Data
A catalogue record for this book is available from the British Library

Library of Congress Cataloging-in-Publication Data
A catalog record for this book is available from the Library of Congress

ISBN–13: 978-0-7506-5805-8
ISBN–10: 0-7506-5805-3

For information on all Newnes publications
visit our website at www.newnespress.com

Transferred to Digital Printing in 2009

Working together to grow
libraries in developing countries

www.elsevier.com | www.bookaid.org | www.sabre.org

ELSEVIER   BOOK AID
            International   Sabre Foundation

Contents

5 Local area network systems 142

Preface

SCADA (or supervisory control and data acquisition) systems are a rapidly growing field with the emphasis being on software and industrial data communications. It should be noted at the outset that the term SCADA has different meanings depending on the industry you are working in. For example, if you are installing a process control system for an industrial plant or factory you will understand SCADA to mean the software operating on a PC which enables you to get a 'window into your process plant' and also to control your plant from a keyboard. On the other hand if you are a water utility engineer and have your water reticulation network scattered over a few thousand square kms with operator stations in a central control room indicating the levels of the various water reservoirs and flow rates in the pipes and activities of the various pumping stations, you will understand SCADA in its traditional role of mainly data acquisition with a small amount of control. In this case the transfer of data will be across radio communications links that are operating over significant distances.

This book covers the fundamentals of SCADA systems hardware, software and the associated communications systems (such as RS-232 and Ethernet) that connect the SCADA operator stations together. It provides you with the tools to understand the basic concepts of SCADA and thus to design and install your next SCADA system more effectively.

After reading this book we hope you will be able to:

- Demonstrate a fundamental understanding of SCADA systems
- Understand the typical SCADA protocols used
- Set up basic industrial networks for SCADA systems
- Grasp some of the basics of troubleshooting the SCADA system hardware

This book is intended for engineers and technicians who are:

- Instrumentation and control engineers
- Process control engineers
- Electrical engineers
- Sales engineers
- Project engineers
- Maintenance supervisors

A basic knowledge of electrical principles is useful in understanding the concepts outlined in the book. But the contents are of a fundamental nature and hence easy to comprehend.

The structure of the book is as follows.

Chapter 1: Background to SCADA. This chapter gives a brief overview of what is covered in the book with an outline of the essentials and a background to SCADA systems.

Chapter 2: SCADA systems, hardware and firmware. The aim of this chapter is to review the typical hardware underpinning SCADA systems with an emphasis on remote terminal units (RTUs) and programmable logic controllers (PLCs).

Chapter 3: SCADA systems software and protocols. The software and associated protocols are the key to understanding SCADA systems and these are examined in some depth in this chapter.

Chapter 4: Landlines. Most SCADA systems have some connection to the telecommunications infrastructure. This chapter assesses the different options open to you in using these different systems.

Chapter 5: Local area network systems. Very few SCADA systems operate in isolation today and depend on a connection to other computers forming a local area network; invariably based on industrial Ethernet. This chapter covers this in some depth.

Chapter 6: Modems. Modems and the ubiquitous RS-232/RS-485 standards are discussed in this chapter. It should be noted that although this is perhaps not as important as in earlier years due to the use of direct digital communications and local area networks, modems and RS-232 are still widely used.

Chapter 7: Central site computer facilities. An oft neglected area, this is examined with a discussion on appropriate installation practises and ergonomic requirements.

Chapter 8: Troubleshooting and maintenance. One of the most interesting areas is encapsulated in this chapter with a discussion on fixing typical RTU problems.

Chapter 9: Specification of systems. This chapter contains a short discussion on specification of SCADA systems and some of the trends in the market.

1

Background to SCADA

1.1   Introduction and brief history of SCADA

This manual is designed to provide a thorough understanding of the fundamental concepts and the practical issues of SCADA systems. Particular emphasis has been placed on the practical aspects of SCADA systems with a view to the future. Formulae and details that can be found in specialized manufacturer manuals have been purposely omitted in favor of concepts and definitions.

This chapter provides an introduction to the fundamental principles and terminology used in the field of SCADA. It is a summary of the main subjects to be covered throughout the manual.

SCADA (supervisory control and data acquisition) has been around as long as there have been control systems. The first 'SCADA' systems utilized data acquisition by means of panels of meters, lights and strip chart recorders. The operator manually operating various control knobs exercised supervisory control. These devices were and still are used to do supervisory control and data acquisition on plants, factories and power generating facilities. The following figure shows a sensor to panel system.

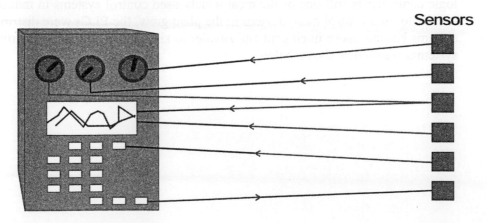

Figure 1.1
Sensors to panel using 4–20 mA or voltage

The sensor to panel type of SCADA system has the following advantages:
- It is simple, no CPUs, RAM, ROM or software programming needed
- The sensors are connected directly to the meters, switches and lights on the panel
- It could be (in most circumstances) easy and cheap to add a simple device like a switch or indicator

The disadvantages of a direct panel to sensor system are:
- The amount of wire becomes unmanageable after the installation of hundreds of sensors
- The quantity and type of data are minimal and rudimentary
- Installation of additional sensors becomes progressively harder as the system grows
- Re-configuration of the system becomes extremely difficult
- Simulation using real data is not possible
- Storage of data is minimal and difficult to manage
- No off site monitoring of data or alarms
- Someone has to watch the dials and meters 24 hours a day

1.2   Fundamental principles of modern SCADA systems

In modern manufacturing and industrial processes, mining industries, public and private utilities, leisure and security industries telemetry is often needed to connect equipment and systems separated by large distances. This can range from a few meters to thousands of kilometers. Telemetry is used to send commands, programs and receives monitoring information from these remote locations.

SCADA refers to the combination of telemetry and data acquisition. SCADA encompasses the collecting of the information, transferring it back to the central site, carrying out any necessary analysis and control and then displaying that information on a number of operator screens or displays. The required control actions are then conveyed back to the process.

In the early days of data acquisition, relay logic was used to control production and plant systems. With the advent of the CPU and other electronic devices, manufacturers incorporated digital electronics into relay logic equipment. The PLC or programmable logic controller is still one of the most widely used control systems in industry. As need to monitor and control more devices in the plant grew, the PLCs were distributed and the systems became more intelligent and smaller in size. PLCs and DCS (distributed control systems) are used as shown below.

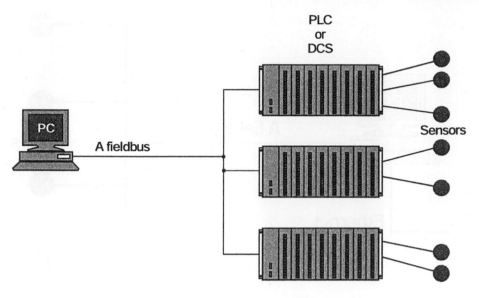

Figure 1.2
PC to PLC or DCS with a fieldbus and sensor

The advantages of the PLC / DCS SCADA system are:
- The computer can record and store a very large amount of data
- The data can be displayed in any way the user requires
- Thousands of sensors over a wide area can be connected to the system
- The operator can incorporate real data simulations into the system
- Many types of data can be collected from the RTUs
- The data can be viewed from anywhere, not just on site

The disadvantages are:
- The system is more complicated than the sensor to panel type
- Different operating skills are required, such as system analysts and programmer
- With thousands of sensors there is still a lot of wire to deal with
- The operator can see only as far as the PLC

As the requirement for smaller and smarter systems grew, sensors were designed with the intelligence of PLCs and DCSs. These devices are known as IEDs (intelligent electronic devices). The IEDs are connected on a fieldbus, such as Profibus, Devicenet or Foundation Fieldbus to the PC. They include enough intelligence to acquire data, communicate to other devices, and hold their part of the overall program. Each of these super smart sensors can have more than one sensor on-board. Typically, an IED could combine an analog input sensor, analog output, PID control, communication system and program memory in one device.

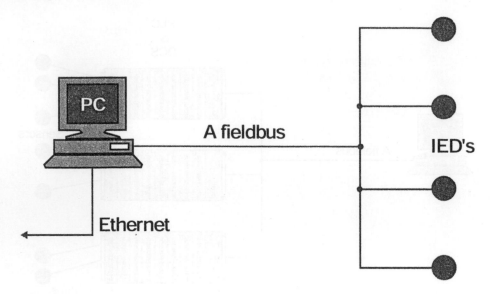

Figure 1.3
PC to IED using a fieldbus

The advantages of the PC to IED fieldbus system are:
- Minimal wiring is needed
- The operator can see down to the sensor level
- The data received from the device can include information such as serial numbers, model numbers, when it was installed and by whom
- All devices are plug and play, so installation and replacement is easy
- Smaller devices means less physical space for the data acquisition system

The disadvantages of a PC to IED system are:
- More sophisticated system requires better trained employees
- Sensor prices are higher (but this is offset somewhat by the lack of PLCs)
- The IEDs rely more on the communication system

1.3   SCADA hardware

A SCADA system consists of a number of remote terminal units (RTUs) collecting field data and sending that data back to a master station, via a communication system. The master station displays the acquired data and allows the operator to perform remote control tasks.

The accurate and timely data allows for optimization of the plant operation and process. Other benefits include more efficient, reliable and most importantly, safer operations. This results in a lower cost of operation compared to earlier non-automated systems.

On a more complex SCADA system there are essentially five levels or hierarchies:
- Field level instrumentation and control devices
- Marshalling terminals and RTUs
- Communications system
- The master station(s)
- The commercial data processing department computer system

The RTU provides an interface to the field analog and digital sensors situated at each remote site.

The communications system provides the pathway for communication between the master station and the remote sites. This communication system can be wire, fiber optic, radio, telephone line, microwave and possibly even satellite. Specific protocols and error detection philosophies are used for efficient and optimum transfer of data.

The master station (or sub-masters) gather data from the various RTUs and generally provide an operator interface for display of information and control of the remote sites. In large telemetry systems, sub-master sites gather information from remote sites and act as a relay back to the control master station.

1.4   SCADA software

SCADA software can be divided into two types, proprietary or open. Companies develop proprietary software to communicate to their hardware. These systems are sold as 'turn key' solutions. The main problem with this system is the overwhelming reliance on the supplier of the system. Open software systems have gained popularity because of the interoperability they bring to the system. Interoperability is the ability to mix different manufacturers' equipment on the same system.

Citect and WonderWare are just two of the open software packages available in the market for SCADA systems. Some packages are now including asset management integrated within the SCADA system. The typical components of a SCADA system are indicated in the next diagram.

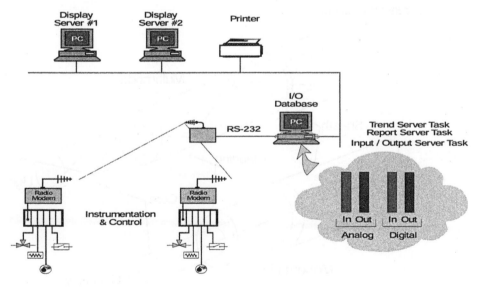

Figure 1.4
Typical SCADA system

Key features of SCADA software are:
- User interface
- Graphics displays
- Alarms
- Trends
- RTU (and PLC) interface
- Scalability

- Access to data
- Database
- Networking
- Fault tolerance and redundancy
- Client/server distributed processing

1.5   Landlines for SCADA

Even with the reduced amount of wire when using a PC to IED system, there is usually a lot of wire in the typical SCADA system. This wire brings its own problems, with the main problem being electrical noise and interference.

Interference and noise are important factors to consider when designing and installing a data communication system, with particular considerations required to avoid electrical interference. Noise can be defined as the random generated undesired signal that corrupts (or interferes with) the original (or desired) signal. This noise can get into the cable or wire in many ways. It is up to the designer to develop a system that will have a minimum of noise from the beginning. Because SCADA systems typically use small voltage they are inherently susceptible to noise.

The use of twisted pair shielded cat5 wire is a requirement on most systems. Using good wire coupled with correct installation techniques ensures the system will be as noise free as possible.

Fiber optic cable is gaining popularity because of its noise immunity. At the moment most installations use glass fibers, but in some industrial areas plastic fibers are increasingly used.

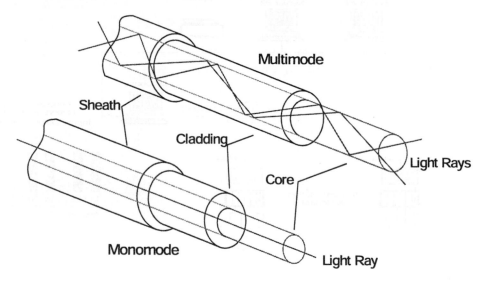

Figure 1.5
Glass fiber optic cables

Future data communications will be divided up between radio, fiber optic and some infrared systems. Wire will be relegated to supplying power and as power requirements of electronics become minimal, even the need for power will be reduced.

1.6 SCADA and local area networks

Local area networks (LAN) are all about sharing information and resources. To enable all the nodes on the SCADA network to share information, they must be connected by some transmission medium. The method of connection is known as the network topology. Nodes need to share this transmission medium in such a way as to allow all nodes access to the medium without disrupting an established sender.

A LAN is a communication path between computers, file-servers, terminals, workstations, and various other intelligent peripheral equipments, which are generally referred to as devices or hosts. A LAN allows access for devices to be shared by several users, with full connectivity between all stations on the network. A LAN is usually owned and administered by a private owner and is located within a localized group of buildings.

Ethernet is the most widely use LAN today because it is cheap and easy to use. Connection of the SCADA network to the LAN allows anyone within the company with the right software and permission, to access the system. Since the data is held in a database, the user can be limited to reading the information. Security issues are obviously a concern, but can be addressed.

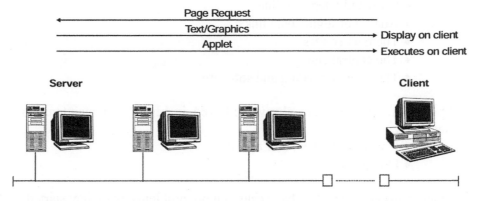

Figure 1.6
Ethernet used to transfer data on a SCADA system

1.7 Modem use in SCADA systems

Figure 1.7
PC to RTU using a modem

Often in SCADA systems the RTU (remote terminal unit (PLC, DCS or IED)) is located at a remote location. This distance can vary from tens of meters to thousands of kilometers. One of the most cost-effective ways of communicating with the RTU over long distances can be by dialup telephone connection. With this system the devices needed are a PC, two dialup modems and the RTU (assuming that the RTU has a built in COM port). The modems are put in the auto-answer mode and the RTU can dial into the PC or the PC can dial the RTU. The software to do this is readily available from RTU manufacturers. The modems can be bought off the shelf at the local computer store.

Line modems are used to connect RTUs to a network over a pair of wires. These systems are usually fairly short (up to 1 kilometer) and use FSK (frequency shift keying) to communicate. Line modems are used to communicate to RTUs when RS-232 or RS-485 communication systems are not practical. The bit rates used in this type of system are usually slow, 1200 to 9600 bps.

1.8 Computer sites and troubleshooting

Computers and RTUs usually run without problems for a long time if left to themselves. Maintenance tasks could include daily, weekly, monthly or annual checks. When maintenance is necessary, the technician or engineer may need to check the following equipment on a regular basis:

- The RTU and component modules
- Analog input modules
- Digital input module
- Interface from RTU to PLC (RS-232/RS-485)
- Privately owned cable
- Switched telephone line
- Analog or digital data links
- The master sites
- The central site
- The operator station and software

Two main rules that are always followed in repair and maintenance of electronic systems are:

- If it is not broken, don't fix it
- Do no harm

Technicians and engineers have caused more problems, than they started with, by doing stupid things like cleaning the equipment because it was slightly dusty. Or trying to get that one more .01 dB of power out of a radio and blown the amplifier in the process.

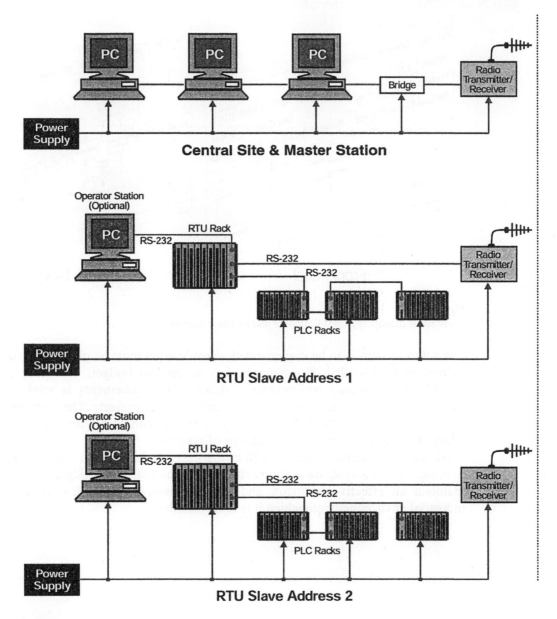

Figure 1.8
Components that could need maintenance in a SCADA system

1.9   System implementation

When first planning and designing a SCADA system, consideration should be given to integrating new SCADA systems into existing communication networks in order to avoid the substantial cost of setting up new infrastructure and communications facilities. This may be carried out through existing LANs, private telephone systems or existing radio systems used for mobile vehicle communications. Careful engineering must be carried out to ensure that overlaying of the SCADA system on to an existing communication network does not degrade or interfere with the existing facilities.

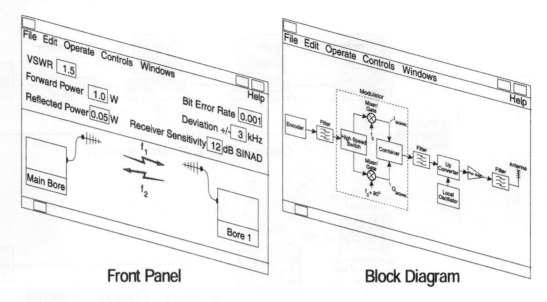

Figure 1.9
Front panel display of SCADA software and its block diagram

If a new system is to be implemented, consideration must be given to the quality of the system to be installed. No company has an endless budget. Weighing up economic considerations against performance and integrity requirements is vital in ensuring a satisfactorily working system at the end of the project. The availability of the communications links and the reliability of the equipment are important considerations when planning performance expectations of systems.

All the aforementioned factors will be discussed in detail in the book. They will then be tied together in a systematic approach to allow the reader to design, specify, install and maintain an effective telemetry and data acquisition system that is suitable for the industrial environment into which it is to be installed.

2

SCADA systems,
hardware and firmware

2.1   Introduction

This chapter introduces the concept of a telemetry system and examines the fundamentals of telemetry systems. The terms SCADA, distributed control system (DCS), programmable logic controller (PLC), smart instrument are defined and placed in the context used in this manual.

The chapter is broken up into the following sections:

- Definitions of the terms SCADA, DCS, PLC and smart instrument
- Remote terminal unit (RTU) structure
- PLCs used as RTUs
- Control site/master station structure
- System reliability and availability
- Communication architectures and philosophies
- Typical considerations in configuration of a master station

The next chapter, which concentrates on the specific details of SCADA systems such as the master station software, communication protocols and other specialized topics will build on the material, contained in this chapter. As discussed in the earlier chapter, the word telemetry refers to the transfer of remote measurement data to a central control station over a communications link. This measurement data is normally collected in real-time (but not necessarily transferred in real-time). The terms SCADA, DCS, PLC and smart instrument are all applications of the telemetry concept.

2.2 Comparison of the terms SCADA, DCS, PLC and smart instrument

2.2.1 SCADA system

A SCADA (or supervisory control and data acquisition) system means a system consisting of a number of remote terminal units (or RTUs) collecting field data connected back to a master station via a communications system. The master station displays the acquired data and also allows the operator to perform remote control tasks.

The accurate and timely data (normally real-time) allows for optimization of the operation of the plant and process. A further benefit is more efficient, reliable and most importantly, safer operations. This all results in a lower cost of operation compared to earlier non-automated systems.

There is a fair degree of confusion between the definition of SCADA systems and process control system. SCADA has the connotation of remote or distant operation. The inevitable question is how far 'remote' is – typically this means over a distance such that the distance between the controlling location and the controlled location is such that direct-wire control is impractical (i.e. a communication link is a critical component of the system).

A successful SCADA installation depends on utilizing proven and reliable technology, with adequate and comprehensive training of all personnel in the operation of the system.

There is a history of unsuccessful SCADA systems – contributing factors to these systems includes inadequate integration of the various components of the system, unnecessary complexity in the system, unreliable hardware and unproven software. Today hardware reliability is less of a problem, but the increasing software complexity is producing new challenges. It should be noted in passing that many operators judge a SCADA system not only by the smooth performance of the RTUs, communication links and the master station (all falling under the umbrella of SCADA system) but also the field devices (both transducers and control devices). The field devices however fall outside the scope of SCADA in this manual and will not be discussed further. A diagram of a typical SCADA system is given opposite.

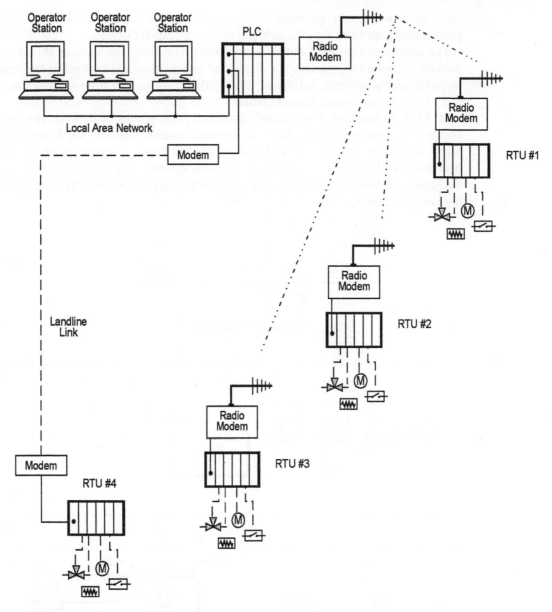

Figure 2.1
Diagram of a typical SCADA system

On a more complex SCADA system there are essentially five levels or hierarchies:

- Field level instrumentation and control devices
- Marshalling terminals and RTUs
- Communications system
- The master station(s)
- The commercial data processing department computer system

The RTU provides an interface to the field analog and digital signals situated at each remote site.

The communications system provides the pathway for communications between the master station and the remote sites. This communication system can be radio, telephone line, microwave and possibly even satellite. Specific protocols and error detection philosophies are used for efficient and optimum transfer of data.

The master station (and submasters) gather data from the various RTUs and generally provide an operator interface for display of information and control of the remote sites. In large telemetry systems, submaster sites gather information from remote sites and act as a relay back to the control master station.

SCADA technology has existed since the early sixties and there are now two other competing approaches possible – distributed control system (DCS) and programmable logic controller (PLC). In addition there has been a growing trend to use smart instruments as a key component in all these systems. Of course, in the real world, the designer will mix and match the four approaches to produce an effective system matching his/her application.

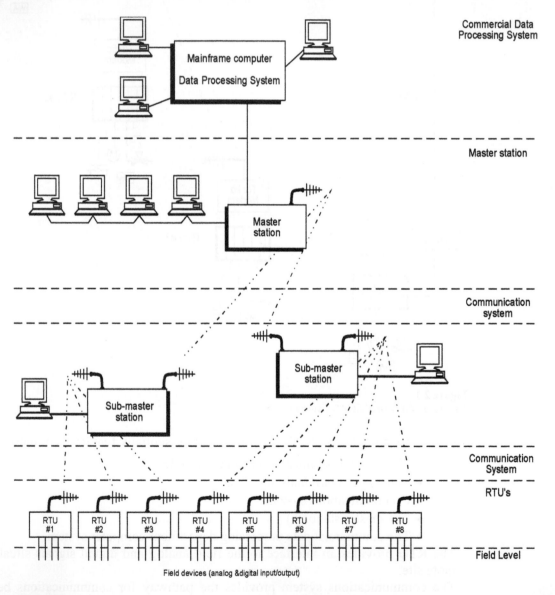

Figure 2.2
SCADA system

2.2.2   Distributed control system (DCS)

In a DCS, the data acquisition and control functions are performed by a number of distributed microprocessor-based units situated near to the devices being controlled or the instrument from which data is being gathered. DCS systems have evolved into systems providing very sophisticated analog (e.g. loop) control capability. A closely integrated set of operator interfaces (or man machine interfaces) is provided to allow for easy system configurations and operator control. The data highway is normally capable of fairly high speeds (typically 1 Mbps up to 10 Mbps).

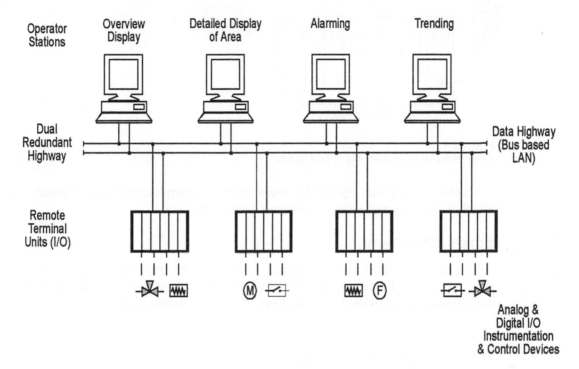

Figure 2.3
Distributed control system (DCS)

2.2.3   Programmable logic controller (PLC)

Since the late 1970s, PLCs have replaced hardwired relays with a combination of ladder–logic software and solid state electronic input and output modules. They are often used in the implementation of a SCADA RTU as they offer a standard hardware solution, which is very economically priced.

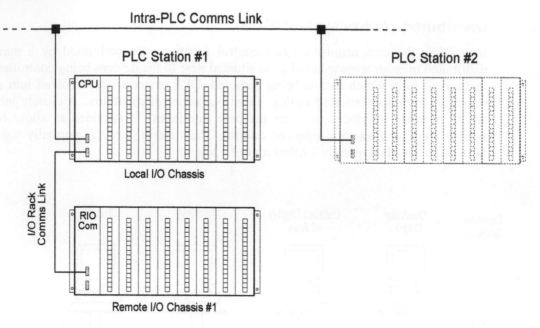

Figure 2.4
Programmable logic controller (PLC) system

Another device that should be mentioned for completeness is the smart instrument which both PLCs and DCS systems can interface to.

2.2.4   Smart instrument

Although this term is sometimes misused, it typically means an intelligent (microprocessor based) digital measuring sensor (such as a flow meter) with digital data communications provided to some diagnostic panel or computer based system.

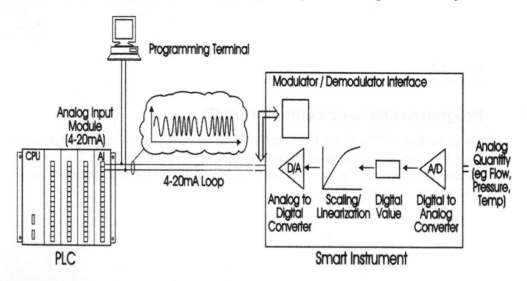

Figure 2.5
Typical example of a smart instrument

This book will henceforth consider DCS, PLC and smart instruments as variations or components of the basic SCADA concept.

2.2.5 Considerations and benefits of SCADA system

Typical considerations when putting a SCADA system together are:

- Overall control requirements
- Sequence logic
- Analog loop control
- Ratio and number of analog to digital points
- Speed of control and data acquisition
- Master/operator control stations
- Type of displays required
- Historical archiving requirements
- System consideration
- Reliability/availability
- Speed of communications/update time/system scan rates
- System redundancy
- Expansion capability
- Application software and modeling

Obviously, a SCADA system's initial cost has to be justified. A few typical reasons for implementing a SCADA system are:

- Improved operation of the plant or process resulting in savings due to optimization of the system
- Increased productivity of the personnel
- Improved safety of the system due to better information and improved control
- Protection of the plant equipment
- Safeguarding the environment from a failure of the system
- Improved energy savings due to optimization of the plant
- Improved and quicker receipt of data so that clients can be invoiced more quickly and accurately
- Government regulations for safety and metering of gas (for royalties & tax etc)

2.3 Remote terminal units

An RTU (sometimes referred to as a remote telemetry unit) as the title implies, is a stand-alone data acquisition and control unit, generally microprocessor based, which monitors and controls equipment at some remote location from the central station. Its primary task is to control and acquire data from process equipment at the remote location and to transfer this data back to a central station. It generally also has the facility for having its configuration and control programs dynamically downloaded from some central station. There is also a facility to be configured locally by some RTU programming unit. Although traditionally the RTU communicates back to some central station, it is also possible to communicate on a peer-to-peer basis with other RTUs. The RTU can also act as a relay station (sometimes referred to as a store and forward station) to another RTU, which may not be accessible from the central station.

Small sized RTUs generally have less than 10 to 20 analog and digital signals, medium sized RTUs have 100 digital and 30 to 40 analog inputs. RTUs, having a capacity greater than this can be classified as large.

A typical RTU configuration is shown in Figure 2.6:

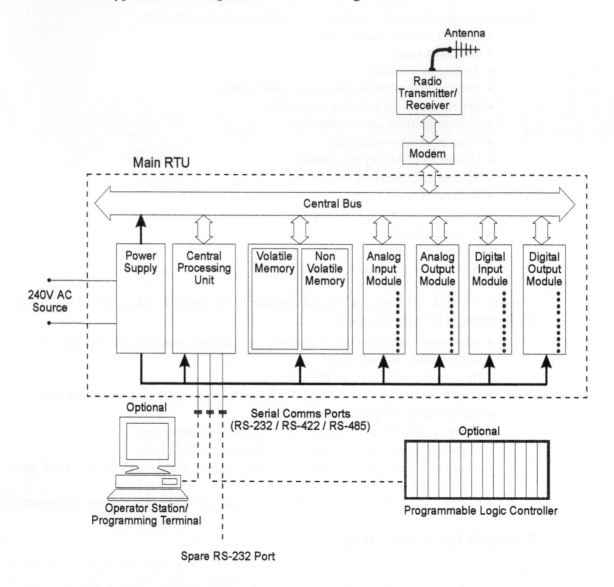

Figure 2.6
Typical RTU hardware structure

A short discussion follows on the individual hardware components.
Typical RTU hardware modules include:

- Control processor and associated memory
- Analog inputs
- Analog outputs
- Counter inputs
- Digital inputs
- Digital outputs

- Communication interface(s)
- Power supply
- RTU rack and enclosure

2.3.1   Control processor (or CPU)

This is generally microprocessor based (16 or 32 bit) e.g. 68302 or 80386. Total memory capacity of 256 kByte (expandable to 4 Mbytes) broken into three types:

1	EPROM (or battery backed RAM)	256 kByte
2	RAM	640 kByte
3	Electrically erasable memory (flash or EEPROM)	128 kByte

A mathematical processor is a useful addition for any complex mathematical calculations. This is sometimes referred to as a coprocessor.

Communication ports – typically two or three ports either RS-232/RS-422/RS-485 for:

- Interface to diagnostics terminal
- Interface to operator station
- Communications link to central site (e.g. by modem)

Diagnostic LEDs provided on the control unit ease troubleshooting and diagnosis of problems (such as CPU failure/failure of I/O module etc).

Another component, which is provided with varying levels of accuracy, is a real-time clock with full calendar (including leap year support). The clock should be updated even during power off periods. The real-time clock is useful for accurate time stamping of events.

A watchdog timer is also required to provide a check that the RTU program is regularly executing. The RTU program regularly resets the watchdog time. If this is not done within a certain time-out period the watchdog timer flags an error condition (and can reset the CPU).

2.3.2   Analog input modules

There are five main components making up an analog input module. They are:

- The input multiplexer
- The input signal amplifier
- The sample and hold circuit
- The A/D converter
- The bus interface and board timing system

A block diagram of a typical analog input module is shown in Figure 2.7.

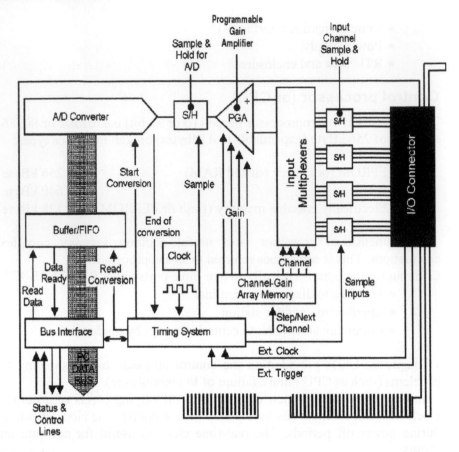

Figure 2.7
Block diagram of a typical analog input module

Each of the individual components will be considered in the following sections.

2.3.2.1 Multiplexers

A multiplexer is a device that samples several (usually 16) analog inputs in turn and switches each to the output in sequence. The output generally goes to an A/D converter, eliminating the need for a converter on each input channel. This can result in considerable savings. A few parameters related to multiplexers are:

- **Crosstalk**
  The amount of signal coupled to the output as a percentage of input signals applied to all OFF channels together.

- **Input leakage current**
  The maximum current that flows into or out of an OFF channel input terminal due to switch leakage.

- **Settling time**
  The time that the multiplexer output takes to settle to a certain percentage (sometimes 90% or sometimes ±1 LSB of the input value) when a single input swings from –FS (full scale) to FS or from +FS to –FS. Essentially, the output must settle to within about ±½ LSB of the input range, before the A/D converter can obtain an accurate conversion of the analog input voltage.

- **Switching time**
  A similar parameter to settling time, it specifies how long the multiplexer output takes to settle to the input voltage when the multiplexer is switched from one channel to another.

- **Throughput rate**
  This relates to the highest rate at which the multiplexer can switch from channel to channel; it is limited by the settling time or the switching time, whichever is longer.

- **Transfer accuracy**
  Expresses the input-to-output error as a percentage of the input.

2.3.2.2 Amplifier

Where low-level voltages need to be digitized, they must be amplified to match the input range of the board's A/D converter. If a low-level signal is fed directly into a board without amplification, a loss of precision will be the result. Some boards provide on-board amplification (or gain), while those with a PGA make it possible to select from software, different gains for different channels, for a series of conversions.

The ideal differential input amplifier only responds to the voltage difference between its two input terminals regardless of what the voltage common to both terminals is doing. Unfortunately, common mode voltages do produce error outputs in real-world amplifiers. An important characteristic is the common mode rejection ratio, CMRR, which is calculated as follows.

$$\text{CMRR} = 20\log\left(V_{\text{cm}} / V_{\text{diff}}\right) \text{ [dB]}$$

where:
V_{cm} *is the voltage common to both inputs*
V_{diff} *is the output (error) voltage when* V_{cm} *is applied to both inputs*
An ideal value for CMRR would be 80 dB or greater.

Drift is another important amplifier specification; it depends on time and temperature. If an amplifier is calibrated to give zero output for zero input at a particular temperature, the output (still at zero input) will change over time and if the temperature changes.

Time drift and temperature drifts are usually measured in PPM/unit time and PPM/°C, respectively. For a 12-bit board, 1 LSB is 1 count in 4096 or 244 PPM. Over an operating range of 0°C to 50°C, a 1 LSB drift is thus:

244 PPM/50°C = 4.88 PPM/°C

In choosing a component, you need to ensure that the board's time and temperature drift specifications over the entire operating temperature range are compatible with the precision you require and don't forget that it can get quite warm inside the RTUs enclosure.

2.3.2.3 Sample-and-hold circuit

Most A/D converters require a fixed time during which the input signal remains constant (the aperture time) in order to perform an A/D conversion. This is a requirement of the conversion algorithm used by the A/D converter. If the input were to change during this time, the A/D would return an inaccurate reading. Therefore, a sample-and-hold device is used on the input to the A/D converter. It samples the output signal from the multiplexer or gain amplifier very quickly and holds it constant for the A/Ds aperture time.

The standard design approach is to place a simple sample-and-hold chip between multiplexer and A/D converter.

2.3.2.4   A/D converters

The A/D converter is the heart of the module. Its function is to measure an input analog voltage and to output a digital code corresponding to the input voltage.

There are two main types of A/D converters used:

- **Integrating (or dual slope) A/Ds**

  These are used for very low frequency applications (a few hundred hertz maximum) and may have very high accuracy and precision (e.g. 22 bit). They are found in thermocouple and RTD modules. Other advantages include very low cost, noise and mains pickup tend to be reduced by the integrating and dual slope nature of the A/D converter. The A/D procedure essentially requires a capacitor to be charged with the input signal for a fixed time, and then uses a counter to calculate how long it takes for the capacitor to discharge. This length of time is proportional to the input voltage.

- **Successive approximation A/Ds**

  Successive approximation A/Ds allow much higher sampling rates (up to a few hundred kHz with 12 bits is possible) while still being reasonable in cost. The conversion algorithm is similar to that of a binary search, where the A/D starts by comparing the input with a voltage (generated by an internal D/A converter), corresponding to half of the full-scale range. If the input is in the lower half, the first digit is zero and the A/D repeats this comparison using the lower half of the input range. If the voltage had been in the upper half, the first digit would have been 1. This dividing of the remaining fraction of the input range in half and comparing to the input voltage continues until the specified number of bits of accuracy have been obtained. It is obviously important that the input signal does not change when the conversion process is underway.

The specifications of A/D converters are discussed below.

- **Absolute accuracy**

  This value refers to the maximum analog error; it is referenced to the national bureau of standards' standard volt.

- **Differential linearity**

  This is the maximum deviation of an actual bit size from its theoretical value for any bit over the full range of the converter.

- **Gain error (scale factor error)**

  The difference in slope between the actual transfer function and the ideal function in percentage.

- **Unipolar offset**

  The first transition should occur ½ LSB above analog common. The unipolar offset error is the deviation of the actual transition point from the ideal first transition point. It is usually adjustable to zero with calibration software and a trimpot on the board. This parameter also usually has an associated temperature drift specification.

- **Bipolar offset**
  Similarly, the transition from FS/2-½ LSB to FS/2 (7 FFh to 800 h on a 12-bit A/D) should occur at ½ LSB below analog common. The bipolar offset (again, usually adjustable with a trimpot) and the temperature coefficient specify the initial deviation and the maximum change in the error over temperature.

- **Linearity errors**
  With most A/D converters gain, offset and zero errors are not critical as they may be calibrated out. Linearity errors, differential non-linearity (DNL) and integral non-linearity INL) are more important because they cannot be removed.

- **Differential non-linearity**
  Is the difference between the actual code width from the ideal width of 1 LSB. If DNL errors are large, the output code widths may represent excessively large and small input voltage ranges. If the magnitude of a DNL is greater than 1 LSB, then at least one code width will vanish, yielding a missing code.

- **Integral non-linearity**
  Is the deviation of the actual transfer function from the ideal straight line. This line may be drawn through the center of the ideal code widths (center-of-code or CC) or through the points where the codes begin to change (low side transition or LST). Most A/Ds are specified by LST INL. Thus the line is drawn from the point ½ LSB on the vertical axis at zero input to the point 1½ LSB beyond the last transition at full-scale input.

- **Resolution**
  This is the smallest change that can be distinguished by an A/D converter. For example, for a 12-bit A/D converter this would be $^{1}/_{4096} = 0.0244\%$.

- **Missing code**
  This occurs when the next output code misses one or more digits from the previous code.

- **Monotonicity**
  This requires a continuously increasing output for a continuously increasing input over the full range of the converter.

- **Quantizing uncertainty**
  Because the A/D can only resolve an input voltage to a finite resolution of   1 LSB, the actual real-world voltage may be up to ½ LSB below the voltage corresponding to the output code or up to ½ LSB above it. An A/D's quantizing uncertainty is therefore always ±½ LSB.

- **Relative accuracy**
  This refers to the input to output error as a fraction of full scale with gain and offset error adjusted to zero.

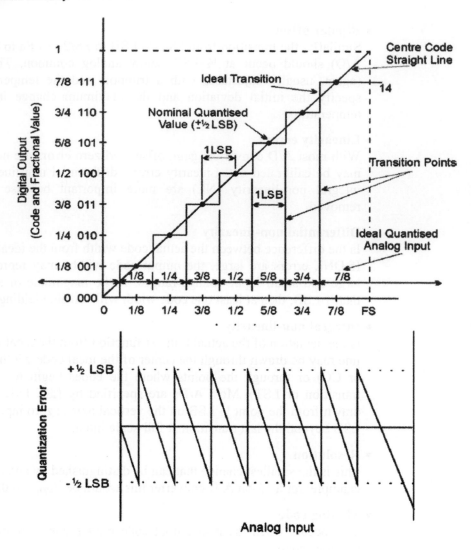

Figure 2.8
Ideal transfer function of an A/D converter with quantization error

The bus interface provides the mechanism for transferring the data from the board and into the host PCs memory, and for sending any configuration information (for example, gain/channel information) or other commands to the board. The interface can be 8-, 16- or 32-bit.

2.3.2.5 Analog input configurations

It is important to take proper care when connecting external transducers or similar devices (the signal source); otherwise the introduction of errors and inaccuracies into a data acquisition system is virtually guaranteed.

2.3.2.6 Connection methods

There are two methods of connecting signal sources to the data acquisition board: Single-ended and differential that are shown below. In general, differential inputs should be used for maximum immunity. Single-ended inputs should only be used where it is impossible to use either of the other two methods.

In the descriptions that follow, these points apply:

- All signals are measured relative to the board's analog ground point, AGND, which is 0 V.
- HI and LO refer to the outputs of a signal source, with LO (sometimes called the signal return) being the source's reference point and HI being the signal value. E_{sn} represents the signal values (that is, VHIn − VLOn) in the diagrams, where *n* is the signal's channel number.
- AMP LO is the reference input of the board's differential amplifier. It is not the same as AGND but it may be referenced to it.
- Because of lead resistance, etc, the remote signal reference point (or ground) is at a different potential to AGND. This is called the common mode voltage V_{CM}. In the ideal situation V_{CM} would be 0 V, but in real-world systems V_{CM} is not 0 V. The voltage at the board's inputs is therefore $E_{sn} + V_{CM}$.

Single-ended inputs

Boards that accept single-ended inputs have a single input wire for each signal, the source's HI side. All the LO sides of the sources are commoned and connected to the analog ground AGND pin. This input type suffers from loss of common mode rejection and is very sensitive to noise. It is not recommended for long leads (longer than ½ m) or for high gains (greater than 5×). The advantage of this method is that it allows the maximum number of inputs, is simple to connect (only one common or ground lead necessary) and it allows for simpler A/D front-end circuitry. We can see from Figure 2.9 that because the amplifier LO (Negative) terminal is connected to AGND, what is amplified is the difference between $E_{sn} + V_{CM}$ and AGND, and this introduces the common mode offset as an error into the readings. Some boards do not have an amplifier, and the multiplexer output is fed straight to the A/D. Single-ended inputs must be used with these types of boards.

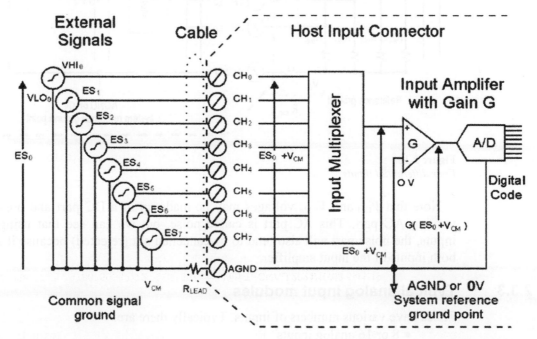

Figure 2.9
Eight single-ended inputs

Differential inputs

True differential inputs provide the maximum noise immunity. This method must also be used where the signal sources have different ground points and cannot be connected together. Referring to Figure 2.10, we see that each channel's individual common mode voltage is fed to the amplifier negative terminal, the individual V_{CMn} voltages are thus subtracted on each reading. Note that two input multiplexers are needed and for the same number of input terminals as single-ended and pseudo-differential inputs, only half the number of input channels is available in differential mode. Also, bias resistors may be required to reference each input channel to ground. This depends on the board's specifications (the book will explain the exact requirements), but it normally consists of one large resistor connected between each signal's LO side and AGND (at the signal end of the cable) and sometimes it requires another resistor of the same value between the HI side and AGND.

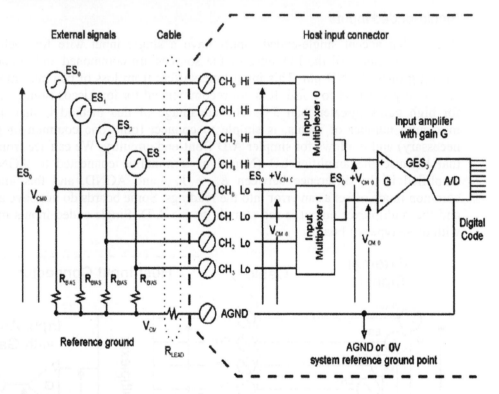

Figure 2.10
Four differential inputs

Note that V_{CM} and V_{CMn} voltages may be made up of a DC part and possibly a time-varying AC part. This AC part is called noise, but we can see that using differential inputs, the noise part will also tend to be cancelled out (rejected) because it is present on both inputs of the input amplifier.

2.3.3 Typical analog input modules

These have various numbers of inputs. Typically there are:
- 8 or 16 analog inputs
- Resolution of 8 or 12 bits
- Range of 4–20 mA (other possibilities are 0–20 mA/±10 volts/0–10 volts)

- Input resistance typically 240 kΩ to 1 MΩ
- Conversion rates typically 10 microseconds to 30 milliseconds
- Inputs are generally single ended (but also differential modes provided)

For reasons of cost and minimization of data transferred over a radio link, a common configuration is eight single ended 8-bit points reading 0–10 volts with a conversion rate of 30 milliseconds per analog point.

An important but often neglected issue with analog input boards is the need for sampling of a signal at the correct frequency. The Nyquist criterion states that a signal must be sampled at a minimum of two times its highest component frequency. Hence the analog to digital system must be capable of sampling at a sufficiently high rate to be well outside the maximum frequency of the input signal. Otherwise filtering must be employed to reduce the input frequency components to an acceptable level. This issue is often neglected due to the increased cost of installing filtering with erroneous results in the measured values. It should be realized the software filtering is not a substitute for an inadequate hardware filtering or sampling rate. This may smooth the signal but it does not reproduce the analog signal faithfully in a digital format.

2.3.4   Analog outputs

2.3.4.1   Typical analogue output module

Typically the analogue output module has the following features:
- 8 analogue outputs
- Resolution of 8 or 12 bits
- Conversion rate from 10 μ seconds to 30 milliseconds
- Outputs ranging from 4–20 mA/± 10 volts/0 to 10 volts

Care has to be taken here on ensuring the load resistance is not lower than specified (typically 50 kΩ) or the voltage drop will be excessive.

Analog output module designs generally prefer to provide voltage outputs rather than current output (unless power is provided externally), as this places lower power requirements on the backplane.

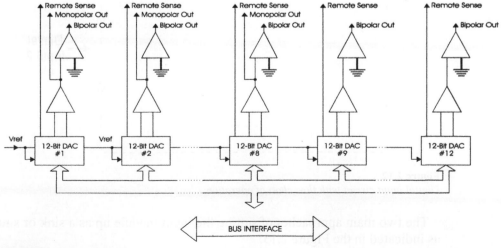

Figure 2.11
Typical analog output module

2.3.5   Digital inputs

These are used to indicate items such as status and alarm signals. Status signals from a valve could comprise two limit switches with contact closed indicating valve - open status and the other contact closed indicating valve – closed status. When both open and closed status contacts are closed, this could indicate the valve is in transit. (There would be a problem if both status switches indicate open conditions.) A high level switch indicates an alarm condition.

It is important with alarm logic that the RTU should be able to distinguish the first alarm from the subsequent spurious alarms that will occur.

Most digital input boards provide groups of 8, 16 or 32 inputs per board. Multiple boards may need to be installed to cope with numerous digital points (where the count of a given board is exceeded).

The standard, normally open or normally closed converter may be used for alarm. In general, normally closed alarm digital inputs are used where the circuit is to indicate an alarm condition.

The input power supply must be appropriately rated for the particular convention used, normally open or normally closed. For the normally open convention, it is possible to de-rate the digital input power supply.

Optical isolation is a good idea to cope with surges induced in the field wiring. A typical circuit and its operation are indicated in Figure 2.12.

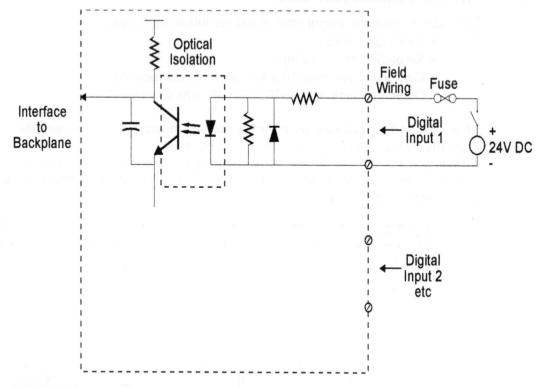

Figure 2.12
Digital input circuit with flow chart of operation

The two main approaches of setting the input module up as a sink or source module are as indicated in the Figure 2.13.

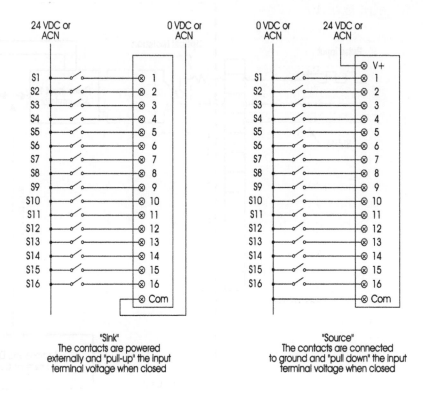

Figure 2.13
Configuring the input module as a sink or source

2.3.5.1   Typical digital input module

Typically the following would be expected of a digital input module:
- 16 digital inputs per module
- Associated LED indicator for each input to indicate current states
- Digital input voltages vary from 110/240 VAC and 12/24/48 VDC
- Optical isolation provided for each digital input

2.3.6   Counter or accumulator digital inputs

There are many applications where a pulse-input module is required – for example from a metering panel. This can be a contact closure signal or if the pulse frequency is high enough, solid state relay signals.

Pulse input signals are normally 'dry contacts' (i.e. the power is provided from the RTU power supply rather than the actual pulse source).

The figure below gives the diagram of the counter digital input system. Optical isolation is useful to minimize the effect of externally generated noise. The size of the accumulator is important when considering the number of pulses that will be counted, before transferring the data to another memory location. For example, a 12-bit register has the capacity for 4096 counts. 16-bit gives 65536 pulses, which could represent 48 minutes @ 20 000 barrels/hour, for example. If these limits are ignored, the classical problem of the accumulator cycling through zero when full could occur.

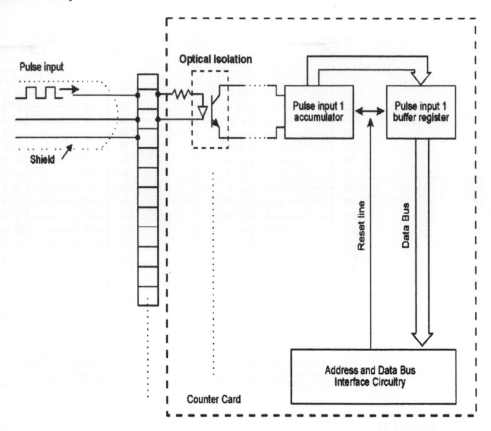

Figure 2.14
Pulse input module

Two approaches are possible:

- The accumulator contents can be transferred to RAM memory at regular intervals where the old and current value difference can be stored in a register.
- The second approach is where a detailed and accurate accounting needs to be made of liquids flowing into and out of a specific area. A freeze accumulator command is broadcast instantaneously to all appropriate RTUs. The pulse accumulator will then freeze the values at this time and transfer to a memory location, and resets the accumulator so that counting can be resumed again.

2.3.6.1 Typical counter specifications

The typical specifications here are:

- 4 counter inputs
- Four 16 bit counters (65 536 counts per counter input)
- Count frequency up to 20 kHz range
- Duty cycle preferably 50% (ratio of mark to space) for the upper count frequency limits.

Note that the duty rating is important, as the counter input needs a finite time to switch on (and then off). If the on pulse is too short, it may be missed although the count frequency is within the specified limits.

A Schmitt trigger gives the preferred input conditioning although a resistor capacitor combination across the counter input can be a cheap way to spread the pulses out.

2.3.7   Digital output module

A digital output module drives an output voltage at each of the appropriate output channels with three approaches possible:

- Triac switching
- Reed relay switching
- TTL voltage outputs

The TRIAC is commonly used for AC switching. A varistor is often connected across the output of the TRIAC to reduce the damaging effect of electrical transients.

Three practical issues should also be observed:

- A TRIAC output switching device does not completely switch on and off but has low and high resistance values. Hence although the TRIAC is switched off it still has some leakage current at the output.
- Surge currents should be of short duration (half a cycle). Any longer will damage the module.
- The manufacturer's continuous current rating should be adhered to. This often refers to individual channels and to the number of channels. There are situations where all the output channels of the module can be used at full rated current capacity. This can exceed the maximum allowable power dissipation for the whole module.

2.3.7.1   Typical digital output modules

- 8 digital outputs
- 240 V AC/24 V DC (0.5 amp to 2.0 amp) outputs
- Associated LED indicator for each output to indicate current status
- Optical isolation or dry relay contact for each output

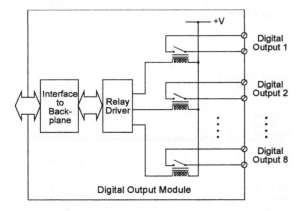

Figure 2.15
Digital output module

'Dry' relay contacts (i.e. no voltage applied to the contacts by the output module) are often provided. These could be reed relay outputs for example. Ensure that the current rating is not exceeded for these devices (especially the inductive current). Although each digital output could be rated at 2 Amps, the module as a whole cannot supply 16 Amps (8 by 2 amps each) and there is normally a maximum current rating for the module of typically 60% of the number of outputs multiplied by the maximum current per output. If this total current is exceeded there will be overheating of the module and eventual failure.

Note also the difference in sinking and sourcing of an I/O module. If a module sinks a specified current, it means that it draws this current from an external source. If a module sources a specific current it drives this current as an output.

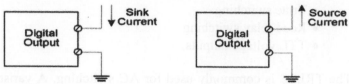

Figure 2.16
Source and sink of current

When connecting to inductive loads it is a good suggestion to put a flywheel diode across the relay for DC systems and a capacitor/resistor combination for AC systems. This minimizes the back EMF effect for DC voltages with the consequent voltage spikes when the devices are switched off.

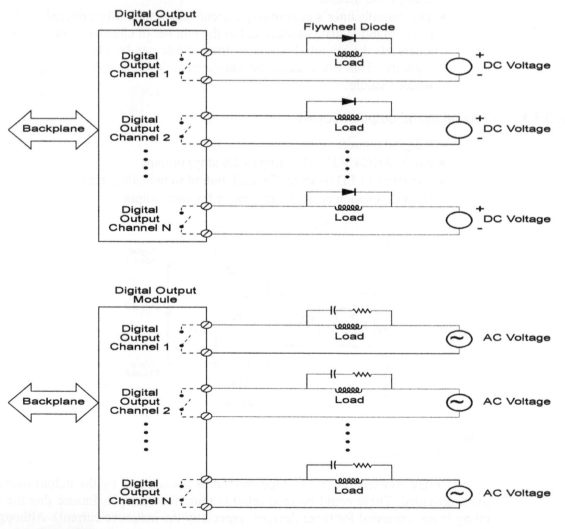

Figure 2.17
Flywheel diode or RC circuit for digital outputs

2.3.8   Mixed analog and digital modules

As many RTUs have only modest requirements, as far as the analog and digital signals are concerned, a typical solution would be to use a mixed analog and digital module. This would typically have:

- 4 analog inputs (8-bit resolution)
- 2 digital inputs
- 1 digital output
- 2 analog output (8-bit resolution)

2.3.9   Communication interfaces

The modern RTU should be flexible enough to handle multiple communication media such as:

- RS-232/RS-442/RS-485
- Dialup telephone lines/dedicated landlines
- Microwave/MUX
- Satellite
- X.25 packet protocols
- Radio via trunked/VHF/UHF/900 MHz

Interestingly enough, the more challenging design for RTUs is the radio communication interface. The landline interface is considered to be an easier design problem. These standards will be discussed in a later section.

2.3.10   Power supply module for RTU

The RTU should be able to operate from 110/240 V AC ± 10% 50 Hz or 12/24/48 V DC ± 10% typically. Batteries that should be provided are lead acid or nickel cadmium. Typical requirements here are for 20-hour standby operation and a recharging time of 12 hours for a fully discharged battery at 25°C. The power supply, battery and associated charger are normally contained in the RTU housing.

Other important monitoring parameters, which should be transmitted back to the central site/master station, are:

- Analog battery reading
- Alarm for battery voltage outside normal range

Cabinets for batteries are normally rated to IP 52 for internal mounting and IP 56 for external mounting.

2.3.11   RTU environmental enclosures

Typically, the printed circuit boards are plugged into a backplane in the RTU cabinet. The RTU cabinet usually accommodates inside an environmental enclosure which protects it from extremes of temperature/weather etc.

Typical considerations in the installations are:

- Circulating air fans and filters: This should be installed at the base of the RTU enclosure to avoid heat buildup. Hot spot areas on the electronic circuitry should be avoided by uniform air circulation. It is important to have a heat soak test too.

- Hazardous areas: RTUs must be installed in explosion proof enclosures (e.g. oil and gas environment).

Typical operating temperatures of RTUs are variable when the RM is located outside the building in a weatherproof enclosure. These temperature specifications can be relaxed if the RTU is situated inside a building, where the temperature variations are not as extreme (provided consideration is given to the situation, where there may be failure of the ventilators or air-conditioning systems).

Typical humidity ranges are 10–95%. Ensure at the high humidity level that there is no possibility of condensation on the circuit boards or there may be contact corrosion or short-circuiting. Lacquering of the printed circuit boards may be an option in these cases. Be aware of the other extreme where low humidity air (5%) can generate static electricity on the circuit boards due to stray capacitance. CMOS based electronics is particularly susceptible to problems in these circumstances. Only screening and grounding the affected electronic areas can reduce static voltages. All maintenance personnel should wear a ground strap on the wrist to minimize the risk of creating and transferring static voltages.

If excessive electromagnetic interference (EMI) and radio frequency interference (RFI) is anticipated in the vicinity of the RTU, special screening and earthing should be used. Some manufacturers warn against using handheld transceivers in the neighborhood of their RTUs. Continuous vibration from vibrating plant and equipment can also have an unfavorable impact on an RTU, in some cases. Vibration shock mounts should be specified for such RTUs. Other areas which should be considered with RTUs are lightning (or protection from electrical surges) and earthquakes (which is equivalent to vibrations at frequencies of 0.1 to 10 Hz).

2.3.12 Testing and maintenance

Many manufacturers provide a test box to test the communications between the RTU and master stations, and also to simulate a master station or RTU in the system. The three typical configurations are indicated below in Figure 2.18.

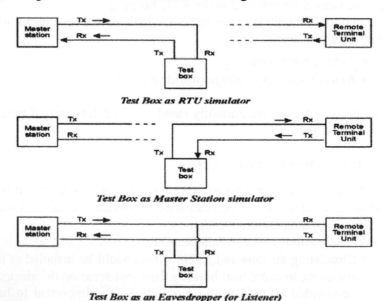

Figure 2.18
SCADA test box operating mode

The typical functions provided on a test box are:

- Message switches: The simulated messages that the user wants to send to the RTU or master station is input here.
- Message indicators: Display of transmit and receiver data.
- Mode of operation: The user selects one of three modes of operation, test box in eavesdropping mode between RTU and master station, test box to RTU, test box to master station. An additional self-test mode is often provided.

There are other features provided such as continuous transmissions of preset messages. Often the test box is interfaced to a PC for easier display and control of actions.

2.3.13 Typical requirements for an RTU system

In the writing of a specification, the following issues should be considered:

Hardware:

Individual RTU expandability (typically up to 200 analog and digital points)
- Off the shelf modules
- Maximum number of RTU sites in a system shall be expandable to 255
- Modular system – no particular order or position in installation (of modules in a rack)
- Robust operation – failure of one module will not affect the performance of other modules
- Minimization of power consumption (CMOS can be an advantage)
- Heat generation minimized
- Rugged and of robust physical construction
- Maximization of noise immunity (due to harsh environment)
- Temperature of –10 to 65°C (operational conditions)
- Relative humidity up to 90%
- Clear indication of diagnostics
- Visible status LEDs
- Local fault diagnosis possible
- Remote fault diagnostics option
- Status of each I/O module and channel (program running/failed/communications OK/failed)
- Modules all connected to one common bus
- Physical interconnection of modules to the bus shall be robust and suitable for use in harsh environments
- Ease of installation of field wiring
- Ease of module replacement
- Removable screw terminals for disconnection and reconnection of wiring

Environmental considerations

The RTU is normally installed in a remote location with fairly harsh environmental conditions. It typically is specified for the following conditions:
- Ambient temperature range of 0 to +60°C (but specifications of –30°C to 60°C are not uncommon)
- Storage temperature range of –20°C to +70°C

- Relative humidity of 0 to 95% non condensing
- Surge withstand capability to withstand power surges typically 2.5 kV, 1 MHz for 2 seconds with 150 ohm source impedance
- Static discharge test where 1.5 cm sparks are discharged at a distance of 30 cm from the unit
- Other requirements include dust, vibration, rain, salt and fog protection.

Software (and firmware)

- Compatibility checks of software configuration of hardware against actual hardware available
- Log kept of all errors that occur in the system both from external events and internal faults
- Remote access of all error logs and status registers
- Software operates continuously despite powering down or up of the system due to loss of power supply or other faults
- Hardware filtering provided on all analog input channels
- Application program resides in non volatile RAM
- Configuration and diagnostic tools for:
- System setup
- Hardware and software setup
- Application code development/management/operation
- Error logs
- Remote and local operation

Each module should have internal software continuously testing the systems I/O and hardware. Diagnostic LEDs should also be provided to identify any faults or to diagnose failure of components. It is important that all these conditions are communicated back to the central station for indication to the operator.

2.4   Application programs

Many applications, which were previously performed at the master station, can now be performed at the RTU, due to improved processing power and memory/dish storage facilities available. Many RTUs also have a local operator interface provided. Typical application programs that can run in the RTU are:

- Analog loop control (e.g. PID)
- Meter proving
- (Gas) flow measurement
- Compressor surge control

2.5   PLCs used as RTUs

A PLC or programmable logic controller is a computer based solid state device that controls industrial equipment and processes. It was initially designed to perform the logic functions executed by relays, drum switches and mechanical timer/counters. Analog control is now a standard part of the PLC operation as well.

The advantage of a PLC over the RTU offerings from various manufacturers is that it can be used in a general-purpose role and can easily be set up for a variety of different functions.

The actual construction of a PLC can vary widely and does not necessarily differ much from generalizing on the discussion of the standard RTU.

PLCs are popular for the following reasons:

- **Economic solution**

  PLCs are a more economic solution than a hardwired relay solution manufactured RTU

- **Versatility and flexibility**

  PLCs can easily have their logic or hardware modified to cope with modified requirements for control

- **Ease of design and installation**

  PLCs have made the design and installation of SCADA systems easier because of the emphasis on software

- **More reliable**

  When correctly installed, PLCs are a far more reliable solution than a traditional hardwired relay solution or short run manufactured RTUs.

- **Sophisticated control**

  PLCs allow for far more sophisticated control (mainly due to the software capability) than RTUs.

- **Physically compact**

  PLCs take up far less space than alternative solutions.

- **Easier troubleshooting and diagnostics**

  Software and clear cut reporting of problems allows easy and swift diagnosis of hardware/firmware/software problems on the system as well as identifying problems with the process and automation system.

A diagram of a PLC and its means of operation using standard ladderlogic are discussed in the following section.

2.5.1   PLC software

The ladder-logic approach to programming is popular because of its perceived similarity to standard electrical circuits. Two vertical lines supplying the power are drawn at each of the sides of the diagram with the lines of logic drawn in horizontal lines.

The example below shows the 'real world' circuit with PLC acting as the control device and the internal ladder-logic within the PLC.

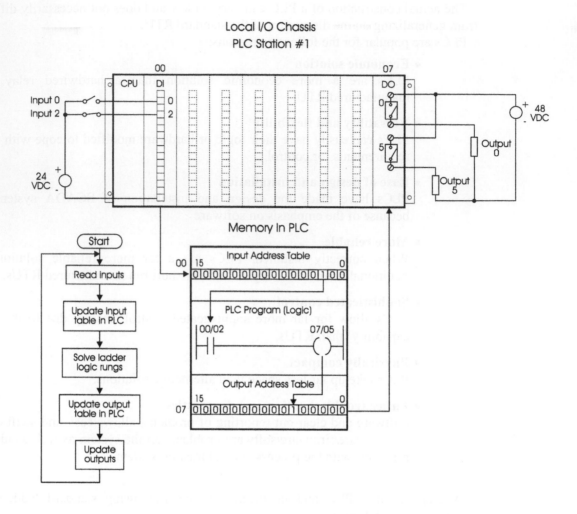

Figure 2.19
The concept of PLC ladder-logic

2.5.2 Basic rules of ladder-logic

The basic rules of ladder-logic can be stated to be:

- The vertical lines indicate the power supply for the control system (12 V DC to 240 V AC). The 'power flow' is visualized to move from left to right.
- Read the ladder diagram from left to right and top to bottom (as in the normal Western convention of reading a book).
- Electrical devices are normally indicated in their normal de-energized condition. This can sometimes be confusing and special care needs to be taken to ensure consistency.
- The contacts associated with coils, timers, counters and other instructions have the same numbering convention as their control device.
- Devices that indicate a start operation for a particular item are normally wired in parallel (so that any of them can start or switch the particular item on). See Figure 2.20 for an example of this.

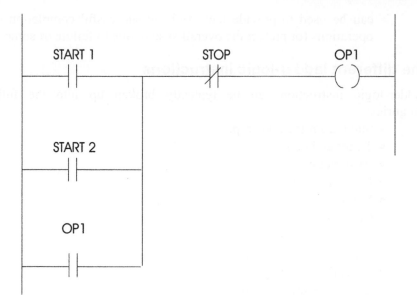

Figure 2.20
Ladder-logic start operation (& logic diagram)

- Devices that indicate a stop operation for a particular item are normally wired in series (so that any of them can stop or switch the particular items off). See Figure 2.21 for an example of this.

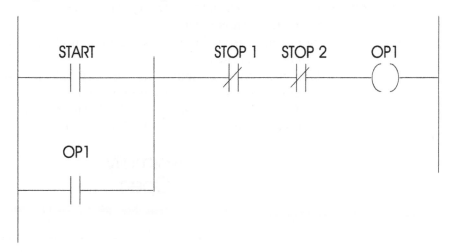

Figure 2.21
Ladder-logic stop operation (& logic diagram)

- Latching operations are used, where a momentary start input signal latches the start signal into the on condition, so that when the start input goes into the OFF condition, the start signal remains energized ON. The latching operation is also referred to as holding or maintaining a sealing contact. See the previous two diagrams for examples of latching.
- Interactive logic: Ladder-logic rungs that appear later in the program often interact with the earlier ladder-logic rungs. This useful feed back mechanism

can be used to provide feed back on successful completion of a sequence of operations (or protect the overall system due to failure of some aspect).

2.5.3 The different ladder-logic instructions

Ladder-logic instruction can be typically broken up into the following different categories:

- Standard relay logic type
- Timer and counters
- Arithmetic
- Logical
- Move
- Comparison
- File manipulation
- Sequencer instructions
- Specialized analog (PID)
- Communication instructions
- Diagnostic
- Miscellaneous (sub routines etc)

A few of these instructions will be discussed in the following sections.

2.5.3.1 Standard relay type

There are three main instructions in this category. These are:

- **Normally open contact**
  (Sometimes also referred to as 'examine if closed' or 'examine on'). The symbol is indicated in Figure 2.22.
  This instruction examines its memory address location for an ON condition. If this memory location is set to ON or 1, the instruction is set to 'ON' or 'TRUE' or '1'. If the location is set to OFF of '0', the instruction is set to 'OFF' or 'FALSE' or '0'.

Figure 2.22
Symbol for normally open contact

- **Normally closed contact**
  (Sometimes also referred to as 'examine if open' or 'examine if off'). This instruction examines its memory address location for an 'OFF' condition. If this memory location is set to OFF of '1', the instruction is set to 'OFF' of '0'. If the memory location is set to ON or '0', the instruction is set to 'ON' or 'TRUE' or '1'. The symbol is indicated in Figure 2.23.

Normally
Closed

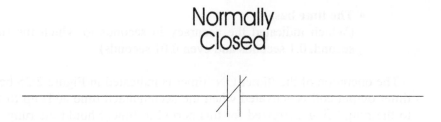

Figure 2.23
Symbol for normally closed contact

2.5.3.2   Output energize coil

When the complete ladder-logic rung is set to a 'TRUE' or 'ON' condition, the output energize instruction sets its memory location to an 'ON' condition; otherwise if the ladder-logic rung is set to a 'FALSE' or 'OFF' condition, the output energize coil sets its memory location to an 'OFF' condition.

The symbol is indicated in Figure 2.24.

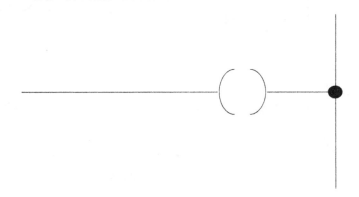

Figure 2.24
Symbol for output energize coil

An example of how the above instructions are used in a practical circuit is indicated diagrammatically in Chapter 3. Note that with coils and contacts, they can refer to 'real world' (or external) or simply internal inputs and outputs.

2.5.3.3   Timers

There are two types of timers:
- Timer ON delay
- Timer OFF delay

There are three parameters associated with each timer:
- **The preset value**
  (Which is the constant number of seconds the timer times to, before being energized or de-energized)
- **The accumulated value**
  (Which is the number of seconds which records how long the timer has been actively timing)

- **The time base**
  (Which indicated the accuracy in seconds to which the timer operates e.g. 1 second, 0.1 seconds and even 0.01 seconds)

The operation of the 'timer ON' timer is indicated in Figure 2.25 below. Essentially the timer output coil is activated when the accumulated time adds up to the preset value due to the rung being energized for this period of time. Should the rung conditions go to the false condition before the accumulator value is equal to the preset value, the accumulator value will be reset to a zero value.

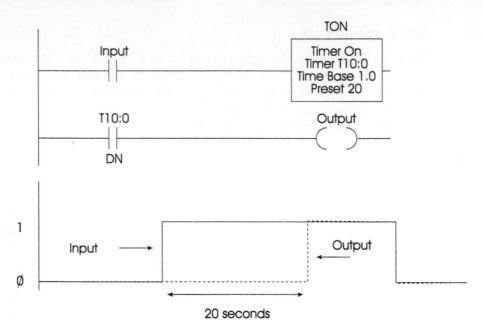

Figure 2.25
Operations of timer on with timing diagram

The operation of the 'timer OFF' timer is that the timer coil is initially energized when the rung is active. As soon as the rung goes false (or inactive) the timer times out (the accumulated value eventually becoming equal to the preset value). At this point the timer coil becomes de-energized. If the rung conditions go low again before the accumulated value reaches the present value, the accumulator is reset to zero. The full sequence of operation is indicated in the Figure 2.26.

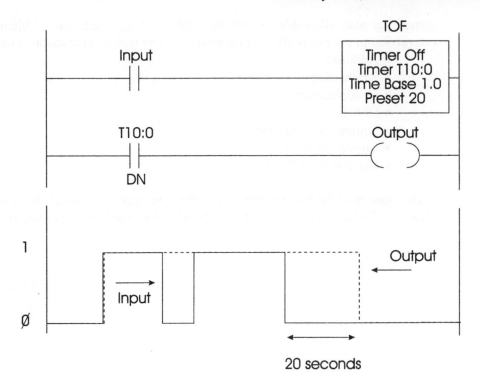

Figure 2.26
Operation of the timer off with timing diagram

2.5.3.4   Counter

There are two types of counters, Count up and Count down. The operation of these counters is very similar to the timer ON and timer OFF timers.

There are two values associated with counters:
- Accumulated value
- Preset value

Count up counters

This counter increments the accumulator value by 1, for every transition of the input contact from false to true. When the accumulated value equals the preset value, the counter output will energize. When a reset instruction is given (at the same address as the counter), the counter is reset and the accumulated value is set to zero.

Count down counters

This counter decrements the accumulator value (which started off at the preset value) by 1, for every transition of the input contact from false to true. When the accumulator value equals zero, the counter output is energized. Interestingly, the counters retain their accumulated count during a power failure, or even if programmed after an MCR instruction.

2.5.3.5   Arithmetic instructions

The various arithmetic instructions are self-explanatory and are generally based around integer floating point arithmetic. The manipulation of an ASCII or BCD value is sometimes also allowable. A full description of manipulation of binary numbers and conversion to integer is given in Appendix C. The typical instructions available are:

- Addition
- Subtraction
- Multiplication
- Division
- Square root extraction
- Convert to BCD
- Convert from BCD

The rung must be true to allow the arithmetic operation (which is situated in the usual location of a rung coil). An example is given for an addition operation in Figure 2.27.

Figure 2.27
Addition operation

Care should be taken when using these operations to monitor control bits such as the carry, overflow, zero and sign bits in case of any problems. The other issue is to ensure that floating point registers are used as destination registers where the source values are floating point, otherwise accuracy will be lost when performing the arithmetic operation.

2.5.3.6 Logical operation

Besides the logical operations that can be performed with relay contacts and coils, which have been discussed earlier, there may be a need to do logical or boolean operations on a 16 bit word.

In the following examples, the bits in equivalent locations of the source words are operated on, bit by bit, to derive the final destination value. The various logical operations that are available are:

- AND
- OR
- XOR (exclusive or)
- NOT (or complement)

The appropriate rung must be true to allow the logical operation (which is situated in the usual location of a rung coil). A full explanation of the meanings of the logical operations is given in Appendix E.

2.5.3.7 Move

This instruction moves the source value at the defined address to the destination address every time this instruction is executed.

Figure 2.28
Move instructions

2.5.3.8 Comparison instructions

These are useful to compare the contents of words with each other. Typical instructions here are to compare two words for:
- Equality
- Not equal
- Less than
- Less than or equal to
- Greater than
- Greater than or equal to

When these conditions are true they can be connected in series with a coil which they then drive into the energized state.

2.5.3.9 Sub routines and jump instructions

There are two main ways of transferring control of the ladder-logic program from the standard sequential path in which it is normally executed. These are:
- Jump to part of the program when a rung condition becomes true (sometimes called jump to a label)
- Jump to a separate block of ladder-logic called a sub routine.

Some users unwittingly run into problems with entry of a ladder-logic rung into the PLC due to limitations in the reporting of incorrect syntax by the relevant packages. The typical limitations are:
- **Numbering of coils and contacts per rung (or network)**
  Most ladder-logic implementations typically allow only one coil per rung, a certain maximum number of parallel branches (e.g. seven) and a certain maximum number of series contacts (e.g. ten) per branch. Additional rungs (with 'dummy' coils) would have to be put in if there was a need for more contacts than can be handled by one rung or network.
- **Vertical contacts**
  Vertical contacts are normally not allowed.
- **Nesting of contacts**
  Contacts may only be allowed to be nested to a certain level in a PLC. In others no nesting is allowed.

- **Direction of power flow**
  'Power flow' within a network or rung always has to be from left to right. Any violation of this principle would be disallowed.

2.6   The master station

The central site/master station can be pictured as having one or more operator stations (tied together with a local area network) connected to a communication system consisting of modem and radio receiver/transmitter. It is possible for a landline system to be used in place of the radio system, in this case the modem will interface directly to the landline. Normally there are no input/output modules connected directly to the master stations although there may be an RTU located in close proximity to the master control room. The features that should be available are:

- Operator interface to display status of the RTUs and enable operator control
- Logging of the data from the RTUs
- Alarming of data from the RTU

As discussed earlier, a master station has two main functions:

- Obtain field data periodically from RTUs and submaster stations
- Control remote devices through the operator station

There are various combinations of systems possible, as indicated in the diagram below.

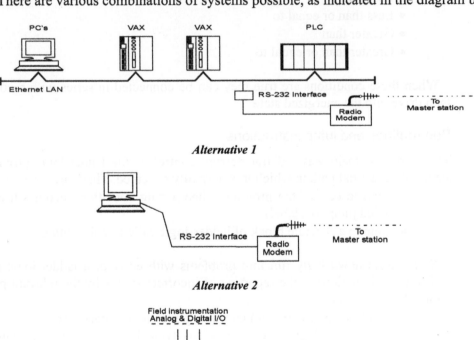

Figure 2.29
Various approaches possible for the master station

It may also be necessary to set up a submaster station. This is to control sites within a specific region. The submaster station has the following functions:

- Acquire data from RTUs within the region
- Log and display this data on a local operator station
- Pass data back to the master station
- Pass on control requests from the master station to the RTUs in its region

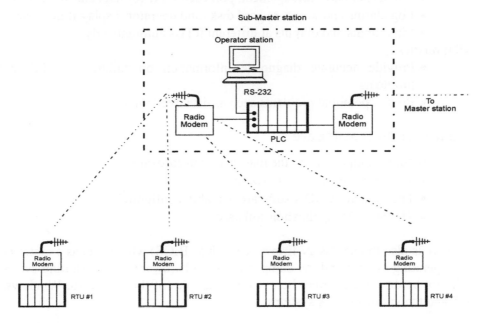

Figure 2.30
Submaster architecture

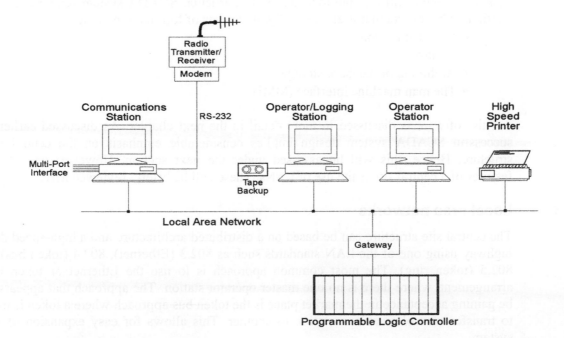

Figure 2.31
Typical structure of the master station

The master station has the following typical functions:

Establishment of communications

- Configure each RTU
- Initialize each RTU with input/output parameters
- Download control and data acquisition programs to the RTU

Operation of the communications link

- If a master slave arrangement, poll each RTU for data and write to RTU
- Log alarms and events to hard disk (and operator display if necessary)
- Link inputs and outputs at different RTUs automatically

Diagnostics

- Provide accurate diagnostic information on failure of RTU and possible problems
- Predict potential problems such as data overloads

2.6.1   Master station software

There are three components to the master station software:

- The operating system software
- The system SCADA software (suitably configured)
- The SCADA application software

There is also the necessary firmware (such as BIOS) which acts as an interface between the operating system and the computer system hardware. The operating system software will not be discussed further here. Good examples of this are DOS, Windows, Windows NT and the various UNIX systems.

2.6.2   System SCADA software

This refers to the software put together by the particular SCADA system vendor and then configured by a particular user. Generally, it consists of four main modules:

- Data acquisition
- Control
- Archiving or database storage
- The man machine interface (MMI)

This software is discussed in more detail in the next chapter. As discussed earlier, a successful SCADA system design implies considerable emphasis on the central site structure. Hence, this will be assessed under the next section. However, one of the features of a central site is the use of LANs. These will be briefly reviewed here.

2.6.3   Local area networks

The central site structure can be based on a distributed architecture and a high-speed data highway using one of the LAN standards such as 802.3 (Ethernet), 802.4 (token bus) or 802.5 (token ring). The most common approach is to use the Ethernet or token bus arrangement, where there is no one master operator station. The approach that appears to be gaining acceptance in the market place is the token bus approach where a token is used to transfer control from one station to another. This allows for easy expansion of the system.

Each of the network options will be discussed in the following paragraphs. Specific reference will be made to the three types of LANs:

- Ethernet (or CSMA/CD)
- Token ring (e.g. IBM token ring)
- Token bus (e.g. MAP/PLC type industrial systems)

Each of these network types is considered in more detail in the following sections.

2.6.4 Ethernet

This is generally implemented as a 10 Mbps baseband coaxial cable network. Carrier sense multiple access with collision detection (or CSMA/CD) is the media access control (or MAC) method used by Ethernet. This is the more popular approach with LANs and hence will be discussed in more detail than the alternative approaches.

The philosophy of Ethernet originated from radio transmission experiments with multiple stations endeavoring to communicate with each other at random times. Essentially before a station (or node) transmits a message (on the common connecting cable) to another node, it first listens for any bus (cable or radio) activity. If it detects that there are no other nodes transmitting, it sends its message. There is a probability that another station may attempt to transmit at precisely the same time. If there is a resultant collision between the two nodes, both nodes will then back off for a random time before reattempting to transmit (hopefully at different times because of the random delay). A typical view of the construction of the medium access control unit for each Ethernet station is given in Figure 2.30.

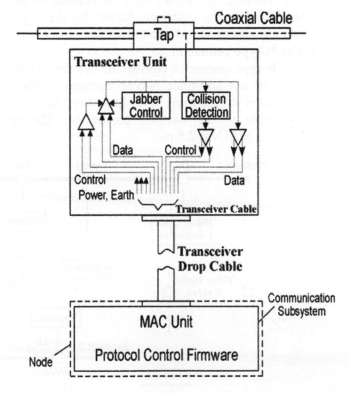

Figure 2.32
A typical hardware layout for a CSMA/CD system

The integrated tap and transceiver unit (referred to as the MAC unit) has the following components:

- A transceiver unit to transmit and receive data, detect collisions, provide electrical isolation and protect the bus from malfunctions
- A tap to make a non intrusive physical connection of the coaxial cable

The controller card, which is connected to the transceiver by shielded cable, contains a medium access control unit, which performs the framing of the messages and error detection, and a microprocessor for implementing the network dependent protocols.

There are three types of Ethernet cabling, standard Ethernet, coaxial Ethernet or 10BASE2 and the 10BASET standard.

Standard Ethernet is referred to in the ISO 8802.3 standard as 10BASE5. This is understood to mean 10 Mb/sec giving baseband transmission with a maximum segment length of 500 m (with each segment having up to 100 MAUs). There is a maximum of five segments allowed in the complete Ethernet system.

The 10BASE5 standard requires 50-ohm coaxial cable with 10.28 mm of that can be tapped using a clamp on tap. This is also referred to as thickwire Ethernet. Male N-connectors are used on the cable for splicing and a female–female N-type connector barrel between the two connectors. The attachment unit interface (AUI) is a 15 conductor shielded cable (consisting of five individually shielded pairs). Each end of a segment must be terminated with a 50 ohm N-connector terminator.

The medium attachment unit (or MAU) is available in two forms:

- **A vampire tap**

  The vampire tap allows easy connection to the coaxial cable as one pin is connected to the center conductor (by drilling a hole into the cable) and the other pin is connected into the shield.

- **The N-type connector**

  The N-type connector consists of two female N connectors thus requiring the cable be cut and male N-type connectors attached where the MAU is to be connected to the trunk line. In a dirty factory environment this approach is preferable to the vampire connection. The minimum connection distance between MAUs is 2.5 m.

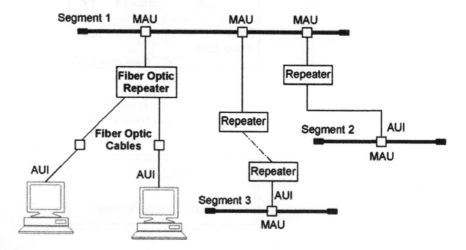

Figure 2.33
Physical layout for 10BASE5 Ethernet

Thin coaxial Ethernet or 10BASE2 were developed after the 10BASE5 standard and is designed to reduce the costs of installation. The maximum segment length is 185 m. Other names for 10BASE2 are Cheapernet and thin wire Ethernet. The coaxial cable used is type RG-58 A/U or C/U with a 50 ohm characteristic impedance.

The trunk coaxial cable of thin Ethernet should not be spliced. Thin Ethernet MAUs may only be connected at consistent distance intervals of 0.5 m. On each 185 m segment of 10BASE2, up to 30 MAUs may be attached, including the MAUs for repeaters. Similar rules are followed as for installation of 10BASE5 except that a segment of thin coaxial cable may not be used as a link segment between two 10BASE5 segments.

A third Ethernet standard is 10BASET. This consists of a star type network with a central hub running twin twisted pair cable to each terminal. The maximum distance a terminal can be from the hub is 100 m. This is for servicing small collections of terminals and is usually tied back into a 10BASE5 network.

A few suggestions on reducing collisions in an Ethernet network are:

- Keep all cables as short as possible
- Keep all high activity sources and their destinations as close as possible. Possibly isolate these nodes from the main network backbone with bridges/routers to reduce backbone traffic.
- Use buffered repeaters rather than bit repeaters
- Check for unnecessary broadcast packets which are aimed at non existent nodes
- Remember that the monitoring equipment to check out network traffic can contribute to the traffic (and the collision rate)
- Ensure that earthing of the cable is done at only one point on one of the cable terminators

2.6.5 Token ring LANs

The second type of network is the token ring system, which was developed by IBM in the early eighties. It is common in the office type environment but not as popular for industrial type systems. It uses a token message to pass control from one node to another. When a node receives the token, it has control of the ring network for a short maximum period of time. When this time is up the token must be passed onto the next node. Alternatively, if the node has no messages to transmit, the token is immediately passed onto the next node.

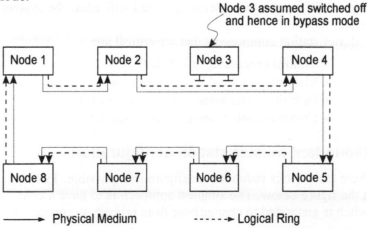

Figure 2.34
Token ring system

Figure 2.34 indicates a typical problem with the ring type network where a node fails (or is switched off) but does not disrupt the operation of the network, due to bypass relays acting to ensure there is a continuous signal path between nodes 2 and 4.

2.6.6   Token bus network

The token bus network is becoming increasingly popular in industrial systems due to its philosophy that all nodes will receive access to a bus with a guaranteed maximum time. The philosophy is similar to that of the token ring network with the use of a token to pass control from node to node.

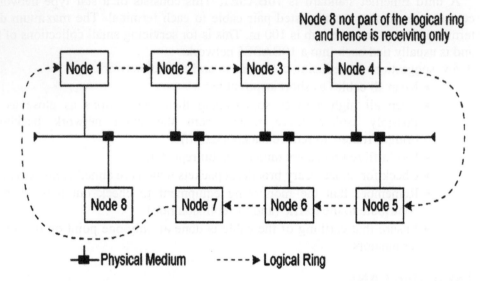

Figure 2.35
Token bus system

2.7   System reliability and availability

The individual component of the SCADA system contributed to the overall reliability of the system. As the master station is a strategic part of the entire SCADA system, it is important that the system reliability and availability are carefully considered. Loss of a single RTU, although unpleasant, should still allow the system to continue to function as before.

Master station components that are critical are:

- Control processing unit (CPU)
- Main memory and buffer reprinters
- Dish drive and associated controller card
- Communications interface and channel

2.7.1   Redundant master station configuration

There are various redundant configurations possible. Two approaches possible are shown in the figure below. The simplest approach is to have a cold standby changeover where a switch is generated to change over from primary to secondary.

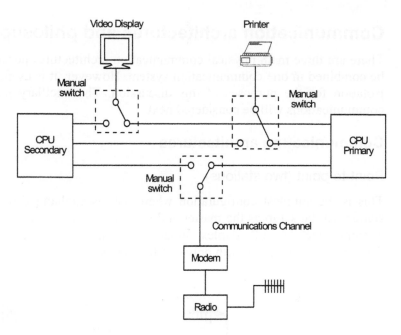

Figure 2.36
Cold standby SCADA system

An example of hot standby configuration is given in the following figure. Here a watchdog timer (WDT) is activated if the primary CPU does not update or reset it within a given time period. Once the WDT is activated, a changeover is effected from the primary to the secondary CPU system. Due to the use of continuous high-speed memory updating of the secondary CPU's memory, the secondary or backup CPU contains all the latest status data (until the WDT activated the changeover).

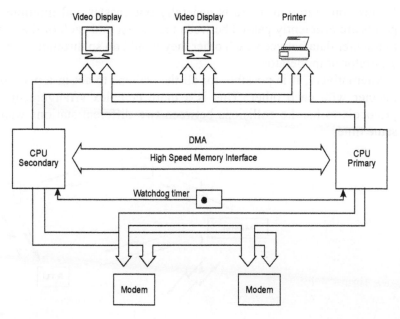

Figure 2.37
Dual ported peripherals

2.8 Communication architectures and philosophies

There are three main physical communication architectures possible. The approaches can be combined in one communication system. However, it is useful to consider each one in isolation for the purposes of this discussion. The ancillary philosophies of achieving communications will be considered next.

2.8.1 Communication architectures

Point-to-point (two stations)

This is the simplest configuration where data is exchanged between two stations. One station can be setup as the master and one as the slave. It is possible for both stations to communicate in full duplex mode (transmitting and receiving on two separate frequencies) or simplex with only one frequency.

Figure 2.38
Point-to-point (two station)

Multipoint (or multiple stations)

In this configuration, there is generally one master and multiple slaves. Generally data points are efficiently passed between the master and each of the slaves. If two slaves need to transfer data between each other they would do so through the master who would act as arbitrator or moderator.

Alternatively, it is possible for all the stations to act in a peer-to-peer communications manner with each other. This is a more complex arrangement requiring sophisticated protocols to handle collisions between two different stations wanting to transmit at the same time.

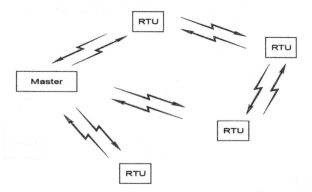

Figure 2.39
Multiple stations

Relay Stations

There are two possibilities here:

- **Store and forward relay operation**

  This can be a component of the other approaches discussed above where one station retransmits messages onto another station out of the range of the master station. This is often called a store and forward relay station. There is no simultaneous transmission of the message by the store and forward station. It retransmits the message at the same frequency as it received it after the message has been received from the master station. This approach is slower than a talk through repeater as each message has to be sent twice. The advantages are considerable savings in mast heights and costs.

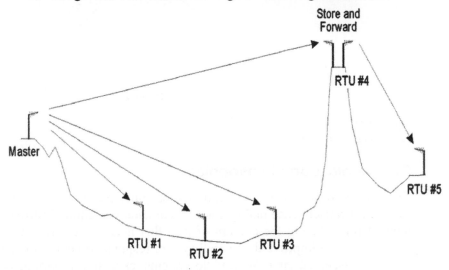

Figure 2.40
Store and forward station

- **Talk through repeaters**

  This is the generally preferred way of increasing the radio system's range. This retransmits a radio signal received simultaneously on another frequency. It is normally situated on a geographically high point. The repeater receives on one frequency and retransmits on another frequency simultaneously. This means that all the stations it is repeating the signal to must receive and transmit on the opposite frequencies. It is important that all stations communicate through the talk through repeater. It must be a common link for all stations and thus have a radio mast high enough to access all RTU sites. It is a strategic link in the communication system; failure would wreak havoc with the entire system. The antenna must receive on one frequency and transmit on a different frequency. This means that the system must be specifically designed for this application with special filters attached to the antennas. There is still a slight time delay in transmission of data with a repeater. The protocol must be designed with this in mind with sufficient lead-time for the repeater's receiver and transmitter to commence operation.

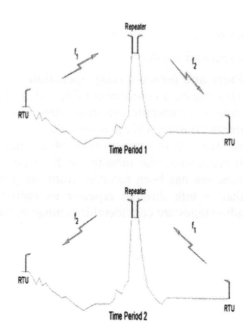

Figure 2.41
Talk through repeaters

2.8.2 Communication philosophies

There are two main communication philosophies possible. These are; polled (or master-slave) and carrier sense multiple access/collision detection (CSMA/CD). The one notable method for reducing the amount of data that needs to be transferred from one point to another (and to improve the overall system response times) is to use exception reporting which is discussed later. With radio systems, exception reporting is normally associated with the CSMA/CD philosophy but there is no theoretical reason why it cannot be applied to RTUs where there is a significant amount of data to be transferred to the master station.

This discussion concentrates on the radio communication aspects. It is difficult to use token bus or CSMA/CD on cable systems other than in a LAN context (with consequent short distances). For longer distances, cable systems would use a polled philosophy.

2.8.3 Polled (or master slave)

This can be used in a point to point or multipoint configuration and is probably the simplest philosophy to use. The master is in total control of the communication system and makes regular (repetitive) requests for data and to transfer data, to and from each one of a number of slaves. The slaves do not initiate the transaction but rely on the master. It is essentially a half-duplex approach where the slave only responds on a request from the master. If a slave does not respond in a defined time, the master then retries (typically up to three times) and then marks the slave as unserviceable and then tries the next slave node in the sequence. It is possible to retry the unserviceable slave again on the next cycle of polling.

The advantages of this approach are:

- Software is easily written and is reliable due to the simplicity of the philosophy.
- Link failure between the master and a slave node is detected fairly quickly.
- No collisions can occur on the network; hence the data throughput is predictable and constant.

- For heavily loaded systems with each node having constant data transfer requirements, as this gives a predictable and efficient system.

The disadvantages are:

- Variations in the data transfer requirements of each slave cannot be handled.
- Interrupt type requests from a slave requesting urgent action cannot be handled (as the master may be processing some other slave).
- Systems, which are lightly loaded with minimum data changes from a slave, are quite inefficient and unnecessarily slow.
- Slaves needing to communicate with each other have to do so through the master with added complexity in the design of the master station.

Two applications of the polled (or master slave) approach are given in the following two implementations. This is possibly the most commonly used technique and is illustrated in the diagram below.

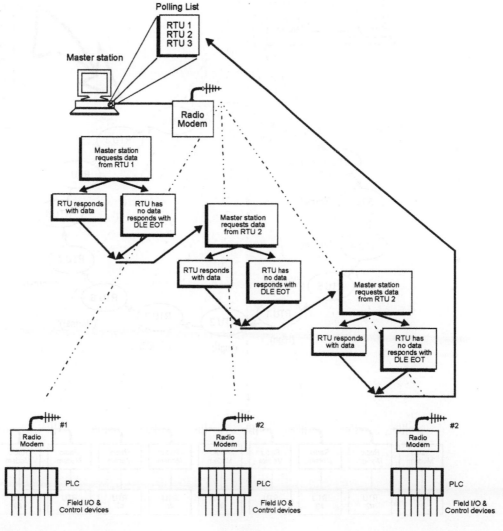

Figure 2.42
Illustration of polling techniques for master station and RTUs

58 *Practical SCADA for Industry*

There are certain considerations to refine the polling scheme beyond what is indicated in the diagram above. These are:

- If there is no response from a given RTU during a poll, a timeout timer has to be set and three retries (in total) initiated before flagging this station as inactive.
- If an RTU is to be treated as a priority station it will be polled at a greater rate than a normal priority station. It is important not to put too many RTUs on the priority list, otherwise the differentiation between high and normal priority becomes meaningless.

An example of a high and normal priority arrangement is given in the diagram below.

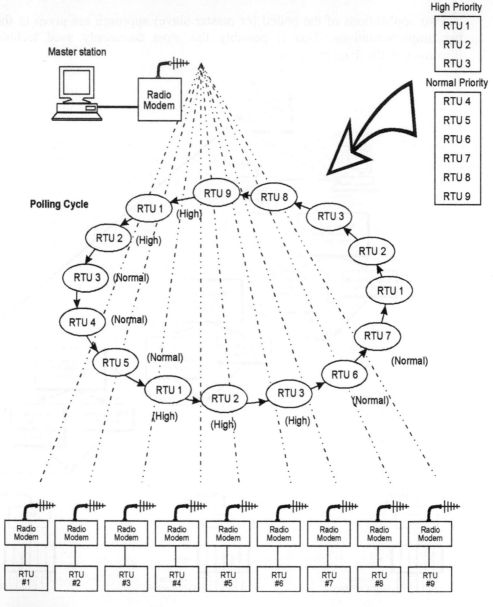

Figure 2.43
High and normal priority arrangement

A priority message sent from the master station can override the standard polling sequence. In this case, the master station completes the poll request for a specific station and then sends out the priority request to a specific station (which is not necessarily next in the polling sequence). It can then wait a predefined time for a response from this RTU or continue with polling a few more stations in the polling sequencer, before requesting a reply from this specific station.

Care should be taken in defining the optimum values for the timers – e.g. a satellite link may have significant development compared to a leased line communications system.

2.8.4   CSMA/CD system (peer-to-peer)

RTU to RTU communication

In a situation where an RTU wants to communicate with another, a solution would be to respond to a poll by the master station having a message with a destination address other than that of the master station's.

The master station will then examine the destination address field of the message received from the RTU and if it does not, mark its own, retransmit onto the appropriate remote station.

This approach can be used in a master slave network or a group of stations all with equal status. It is similar to the operation of Ethernet discussed in section 2.4.1.

The only attempt to avoid collisions is to listen to the medium before transmitting. The systems rely on recovery methods to handle collision problems. Typically these systems are very effective at low capacity rates, as soon as the traffic rises to over 30% of the channel capacity there is an avalanche collapse of the system and communications become unreliable and erratic. The initial experiments with radio transmission between multiple stations (on a peer to peer basis) used CSMA/CD.

This technique is used solely on networks where all nodes have access to the same media (within radio range or on a common cable link). All data is transmitted by the transmitting node first encapsulating the data in a frame with the required destination node address at the head of the frame. All nodes will read this frame and the node which identifies its address at the head of the frame will then continue reading the data and respond appropriately.

However with this style of operation it is possible for two nodes to try and transmit at the same time, with a resultant collision. In order to minimize the chance of a collision, the source node first listens for a carrier signal (indicating that a frame is being transmitted) before commencing transmission. Unfortunately this does not always work where certain stations (which cannot hear each other) try and transmit back to the station simultaneously.

There is a collision here, which only the master can detect (and thus correct). However it is possible that two (or more) transmitting nodes may determine that there is no activity on the system and both start to transmit at the same time. Intuitively, this means that two bits of the same polarity will add together, and the resultant signal seen by the transceivers exceeds that which could come from a single station. A collision is said to occur. The two or more transmitting nodes that were involved in the collision then wait for a further short random time interval before trying to retransmit again.

It is possible (especially on the standard cable type systems) for the transmitting nodes to see a collision when it occurs (with TTR radios) and to enforce the collision by sending a random bit pattern for a short period (called a jam sequence). This would occur before waiting for the random time interval. It ensures that the master site sees the collision.

Exception reporting (or event reporting)

A technique to reduce the unnecessary transfer of data is to use some form of exception reporting. This approach is popular with the CSMA/CD philosophy but it could also offer a solution for the polled approach where there is a considerable amount of data to transfer from each slave.

The remote station monitors its own inputs for a change of state or data. When there is a change of state, the remote station writes a block of data to the master station when the master station polls the remote.

Typical reasons for using polled report by exception include:

- The communications channel is operating at a low data rate (say 4800 bps)
- There is substantial data being monitored at the remote stations (say 80 bits or more)
- There are more than 10 RTUs linked to one master station

Each analog or digital point that reports back to the central master station has a set of exception reporting parameters associated with it. The type of exception reporting depends on the particular environment but could be:

- High and low alarm limits of analog value
- Percent of change in the full span of the analog signal
- Minimum and maximum reporting time intervals

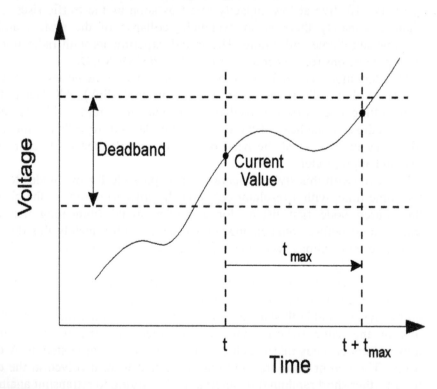

Figure 2.44
Exception reporting system

When an analog value changes in excess of a given parameter or an alarm occurs an exception report is generated. A digital point generates an exception report when the point changes state (from a '0' to a '1' or vice versa).

The advantages of this approach are quite clearly to minimize unnecessary (repetitive) traffic from the communications system.

The disadvantages are essentially:

- The master station may only detect a link failure after a period of time due to the infrequency of communication.
- The data in the system is not always the latest and may be up to 30 minutes old for example.
- There is effectively a filtering action on analog values by the master station, as small variations do not get reported once the analog values are outside the limits.
- The operator must manually institute a system update to gain the latest data from the RTUs.

Polling plus CSMA/CD with exception reporting

A practical and yet novel approach to combining all the approaches discussed previously is to use the concept of a slot time for each station. Assume that the architecture is for a master and a number of slaves, which need to communicate with the master station. There is no communication between slaves required (except possibly through the master).

The time each station is allowed to transmit is called a slot time. There are two types of slots:

- A slave (or a few slaves) transmitting to a master
- A master transmitting to a slave

A slot time is calculated as the sum of the maximums of modem up time (30 milliseconds), plus radio transmit time (100 milliseconds), plus time for protocol message (58.3 milliseconds), plus muting time (25 milliseconds) of transmitter. Typical times are given in brackets after the description.

The master commences operations by polling each slave in turn (and thereafter every 3600 seconds say). Each slave will synchronize in on the master message and will transmit an acknowledged message. The time slots will alternate for the master transmitting and the master receiving. Hence, on a change in state of a slave node it will transmit the data on the first master receiver time slot. If two remote slaves try to transmit in the same time slot, the message will be corrupted and the slaves will not receive a response from the master. The slaves will then select a random master receiver time slot to attempt a retransmission of the message. If the master continues to get corrupted messages, it may elect to do a complete poll of all the remote slaves (as the CSMA/CD type mechanism is possibly breaking down due to excessive traffic).

2.9 Typical considerations in configuring a master station

Before commencing with a detailed discussion, a few factors to bear in mind when designing the system are:

- Simplicity (the 'KISS' principle)
- Minimum response time
- Deterministic type operation (especially for critical signals from RTUs)

- Minimum cost
- Optimum efficiency of operation
- Data format and communication speed (baud rate/stop bits/parity)

While it is difficult to generalize all master stations/RTUs and communication systems forming part of a SCADA system, a few considerations are discussed below:

- **Hardware handshaking**
  If a modem is not being used, select 'no handshaking'. If a modem is being used a full or half-duplex one should be considered.

- **Station address**
  Every station (and RTU) must have a unique address.

- **Error detection**
  Block check (BCC) or cyclic redundancy check (CRC). Preferably select CRC as this gives the best error checking capability. This is discussed later in this book.

- **Protocol message retries**
  How many retries or messages transmitted before the master station flags this RTU as unavailable? (Typically three retries are standard. It must be noted however that certain sensitive units may require just one retry before the master station flags it as unavailable.)

- **RTS send delay**
  This typically goes with the clear to send signal from the modem. This defines the amount of time that elapses before the message is transmitted. The clear to send line (from the modem) must also be asserted before transmission can occur.

- **RTS off delay**
  This is the time that elapses between the end of the message and the inhibiting of the RTS signal. It is important not to intermit the message prematurely by setting this RTS off delay, too short.

- **Timeout delay**
  The timeout delay for a message received from an RTU device.

- **Size of messages from RTU**
  This defines the maximum size of messages allowable from the RTU during a poll by the master station.

- **Priority message transmit**
  This defines when an immediate message is required to be transmitted by the master station, at the conclusion of a poll of an RTU station or when the master station appears in the poll sequence.

- **Poll sequence**
  Define the station addresses in the poll sequence for both priority and normal message transfers.

- **Addressing considerations**

  Each station on a network should have a unique address. One address (normally FF_{16} or 1111 1111) is reserved for broadcast address. Some protocols also reserve certain of the upper addresses for diagnostic purposes and they should also not be used.

3

SCADA systems, software and protocols

3.1   Introduction

This chapter will focus specifically on SCADA systems and protocol with most emphasis placed on man-machine software.

The following points will be discussed in detail:
- The components of a SCADA system
- The SCADA software package
- Specialized SCADA protocols
- Error detection
- New technologies in SCADA systems
- The twelve golden rules.

3.2   The components of a SCADA system

The typical components of a SCADA system with emphasis on the SCADA software are indicated in the diagram below.

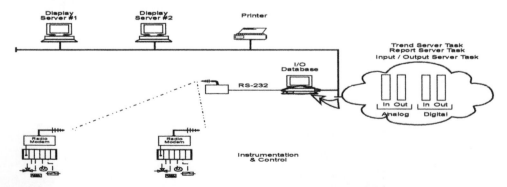

Figure 3.1
Components of a SCADA system

Typical key features expected of the SCADA software are listed below. Naturally these features depend on the hardware to be implemented.

3.2.1   SCADA key features

User interface

- Keyboard
- Mouse
- Trackball
- Touch screen

Graphics displays

- Customer-configurable, object orientated and bit mapped
- Unlimited number of pages
- Resolution: up to 1280 × 1024 with millions of colors

Alarms

- Client server architecture
- Time stamped alarms to 1 millisecond precision (or better)
- Single network acknowledgment and control of alarms
- Alarms are shared to all clients
- Alarms displayed in chronological order
- Dynamic allocation of alarm pages
- User-defined formats and colors
- Up to four adjustable trip points for each analog alarm
- Deviation and rate of change monitoring for analog alarms
- Selective display of alarms by category (256 categories)
- Historical alarm and event logging
- Context-sensitive help
- On-line alarm disable and threshold modification
- Event-triggered alarms
- Alarm-triggered reports
- Operator comments can be attached to alarms

Trends

- Client server architecture
- True trend printouts not screen dumps
- Rubber band trend zooming
- Export data to DBF, CSV files
- X/Y plot capability
- Event based trends
- Pop-up trend display
- Trend gridlines or profiles
- Background trend graphics
- Real-time multi-pen trending

- Short and long term trend display
- Length of data storage and frequency of monitoring can be specified on a per-point basis
- Archiving of historical trend data
- On-line change of time-base without loss of data
- On-line retrieval of archived historical trend data
- Exact value and time can be displayed
- Trend data can be graphically represented in real-time

RTU (and PLC) interface

- All compatible protocols included as standard
- DDE drivers supported
- Interface also possible for RTUs, loop controllers, bar code readers and other equipment
- Driver toolkit available
- Operates on a demand basis instead of the conventional predefined scan method
- Optimization of block data requests to PLCs
- Rationalization of network user data requests
- Maximization of PLC highway bandwidth

Scalability

- Additional hardware can be added without replacing or modifying existing equipment
- Limited only by the PLC architecture (typically 300 to 40 000 points)

Access to data

- Direct, real-time access to data by any network user
- Third-party access to real-time data, e.g. Lotus 123 and Excel
- Network DDE
- DDE compatibility: read, write and exec
- DDE to all IO device points
- Clipboard

Database

- ODBC driver support
- Direct SQL commands or high level reporting

Networking

- Supports all NetBIOS compatible networks such as NetWare, LAN Manager, Windows for Workgroups, Windows NT (changed from existing)
- Support protocols NetBEUI, IPX/SPX, TCP/IP and more
- Centralized alarm, trend and report processing – data available from anywhere in the network
- Dual networks for full LAN redundancy
- No network configuration required (transparent)

- May be enabled via single check box, no configuration
- LAN licensing is based on the number of users logged onto the network, not the number of nodes on the network
- No file server required
- Multi-user system, full communication between operators
- RAS and WAN supported with high performance
- PSTN dial up support

Fault tolerance and redundancy

- Dual networks for full LAN redundancy
- Redundancy can be applied to specific hardware
- Supports primary and secondary equipment configurations
- Intelligent redundancy allows secondary equipment to contribute to processing load
- Automatic changeover and recovery
- Redundant writes to PLCs with no configuration
- Mirrored disk I/O devices
- Mirrored alarm servers
- Mirrored trend servers
- File server redundancy
- No configuration required, may be enabled via single check box, no configuration

Client/server distributed processing

- Open architecture design
- Real-time multitasking
- Client/server fully supported with no user configuration
- Distributed project updates (changes reflected across network)
- Concurrent support of multiple display nodes
- Access any tag from any node
- Access any data (trend, alarm, report) from any node

3.3   The SCADA software package

While performance and efficiency of the SCADA package with the current plant is important, the package should be easily upgradeable to handle future requirement. The system must be easily modifiable as the requirement change and expandable as the task grows, in other words the system must use a scaleable architecture.

There have been two main approaches to follow in designing the SCADA system in the past. They are centralized and distributed.

Centralized, where a single computer or mainframe performs all plant monitoring and all plant data is stored on one database that resides on this computer. The disadvantages with this approach are simply:

- Initial up front costs are fairly high for a small system
- A gradual (incremental) approach to plant upgrading is not really possible due to the fixed size of the system
- Redundancy is expensive as the entire system must be duplicated

- The skills required of maintenance staff in working with a mainframe type computer can be fairly high

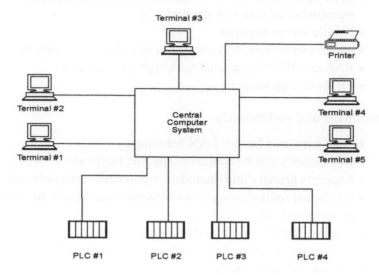

Figure 3.2
Centralized processing

Distributed: where the SCADA system is shared across several small computers (usually PCs). Although the disadvantages of the centralized approach above are addressed with a distributed system, the problems are:

- Communication between different computers is not easy, resulting in configuration problems
- Data processing and databases have to be duplicated across all computers in the system, resulting in low efficiencies
- There is no systematic approach to acquiring data from the plant devices – if two operators require the same data, the RTU is interrogated twice

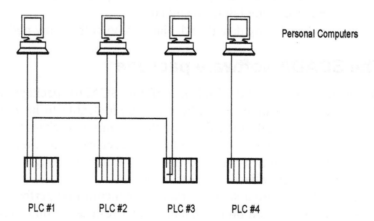

Figure 3.3
Distributed processing

An effective solution is to examine the type of data required for each task and then to structure the system appropriately. A client server approach also makes for a more effective system.

A client server system is understood as follows:

A server node is a device that provides a service to other nodes on the network. A common example of this is a database program. A client on the other hand is a node that requests a service from a server. The word client and server refer to the program executing on a particular node.

A good example is a display system requiring display data. The display node (or client) requests the data from the control server. The control server then searches the database and returns the data requested, thus reducing the network overhead compared to the alternative approach of the display node having to do the database search itself.

A typical implementation of a SCADA system is shown in the figure below.

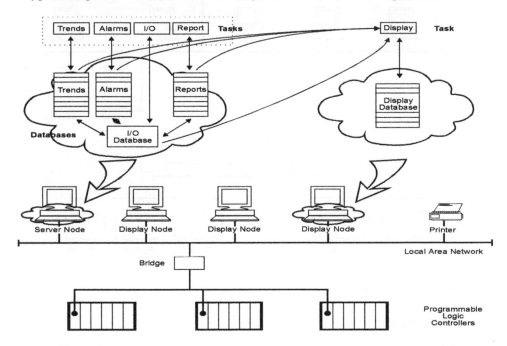

Figure 3.4
Client server approach as applied to a SCADA system

There are typically five tasks in any SCADA system. Each of these tasks performs its own separate processing.

- **Input/output task**
  This program is the interface between the control and monitoring system and the plant floor.
- **Alarm task**
  This manages all alarms by detecting digital alarm points and comparing the values of analog alarm points to alarm thresholds.
- **Trends task**
  The trends task collects data to be monitored over time.
- **Reports task**
  Reports are produced from plant data. These reports can be periodic, event triggered or activated by the operator.
- **Display task**
  This manages all data to be monitored by the operator and all control actions requested by the operator.

A large system with 30 000 points could have architecture as indicated below.

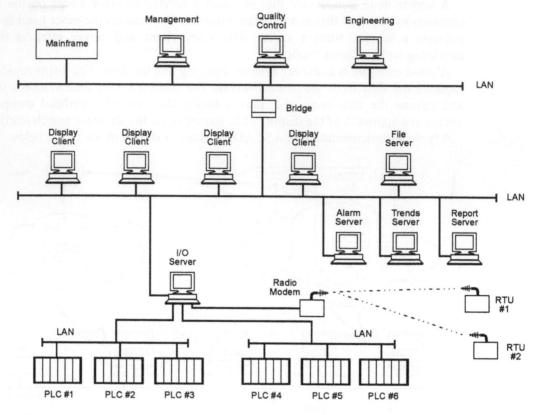

Figure 3.5
A large SCADA application

3.3.1 Redundancy

A typical example of a SCADA system where one component could disrupt the operation of the entire system is given in the diagram below.

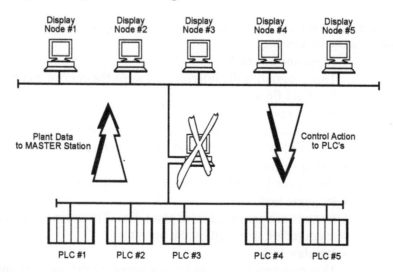

Figure 3.6
The weak link

If any processes or activities in the system are critical, or if the cost of loss of production is high, redundancy must be built into the system.

This can be done in a number of ways as indicated in the following diagrams. The key to the approach is to use the client–server approach, which allows for different tasks (comprising the SCADA system) to run on different PC nodes. For example, if the trend task were important, this would be put in both the primary and secondary servers.

The primary server would constantly communicate with the secondary server updating its status and the appropriate databases. If the primary server fails, the standby server will then take over as the primary server and transfer information to the clients on the network.

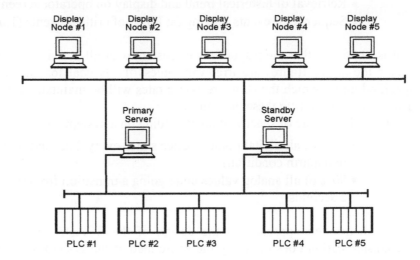

Figure 3.7
Dual server redundancy

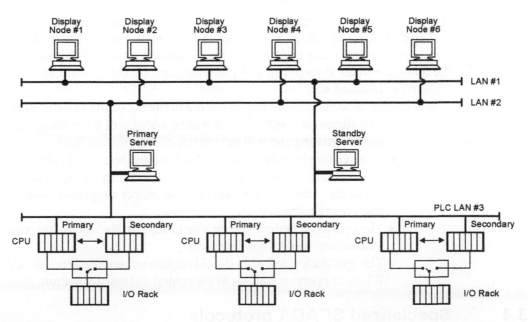

Figure 3.8
Dual LANs and PLCs

3.3.2 System response time

These should be carefully specified for the following events. Typical speeds that are considered acceptable are:

- Display of analog or digital value (acquired from RTU) on the master station operator display (1 to 2 seconds maximum)
- Control request from operator to RTU (1 second critical; 3 seconds non-critical)
- Acknowledge of alarm on operator screen (1 second)
- Display of entire new display on operator screen (1 second)
- Retrieval of historical trend and display on operator screen (2 seconds)
- Sequence of events logging (at RTU) of critical events (1 millisecond)

It is important that the response is consistent over all activities of the SCADA system. Hence the above figures are irrelevant unless the typical loading of the system is also specified under which the above response rates will be maintained. In addition no loss of data must occur during these peak times.

A typical example of specification of loading on a system would be:

- 90% of all digital points change status every 2 seconds (or go from healthy into alarm condition).
- 80% of all analog values undergoing a transition from 0 to 100% every 2 seconds.

The distributed approach to the design of the SCADA system (where the central site/master station does not carry the entire load of the system) ensures that these figures can be easily achieved.

3.3.3 Expandability of the system

A typical figure quoted in industry is that if expansion of the SCADA system is anticipated over the life of the system the current requirements of the SCADA system should not require more than 60% of the processing power of the master station and that the available mass storage (on disk) and memory (RAM) should also be approximately 50% of the required size.

It is important to specify the expansion requirements of the system, so that;

- The additional hardware that will be added will be of the same modular form, as that existing and will not impact on the existing hardware installed.
- The existing installation of SCADA hardware/control cabinets/operator displays will not be unfavorably impacted on by the addition of additional hardware. This includes items such as power supply/air conditioning/SCADA display organization.
- The operating system will be able to support the additional requirements without any major modifications.
- The application software should require no modifications in adding the new RTUs or operator stations at the central site/master station.

3.4 Specialized SCADA protocols

A protocol controls the message format common to all devices on a network. Common protocols used in radio communications and telemetry systems include the HDLC,

MPT1317 and Modbus protocols. The CSMA/CD protocol format is also used and this is discussed in Section 2.6. This section will provide an introduction to protocols and also provide a description of a common protocol used in telemetry, the HDLC protocol.

3.4.1 Introduction to protocols

The transmission of information (both directions) between the master station and RTUs using time division multiplexing techniques requires the use of serial digital messages. These messages must be efficient, secure, flexible, and easily implemented in hardware and software. Efficiency is defined as:

Information bits transmitted ÷ Total bits transmitted

Security is the ability to detect errors in the original information transmitted, caused by noise on the communication channel. Flexibility allows different amounts and types of information to be transmitted upon command by the master station. Implementation in hardware and software requires the minimum in complicated logic, memory storage, and speed of operation.

All messages are divided into three basic parts as follows:
- Message establishment: This provides the signals to synchronize the receiver and transmitter.
- Information: This provides the data in a coded form to allow the receiver to decode the information and properly utilize it.
- Message termination: This provides the message security checks and a means of denoting the end of the message. Message security checks consist of logical operations on the data, which result in a predefined number of check bits transmitted with the message. At the receiver the same operations are performed on the data and compared with the received check bits. If they are identical, the message is accepted; otherwise, a retransmission of the original message is requested.

A typical example of commonly used asynchronous message format is shown below:

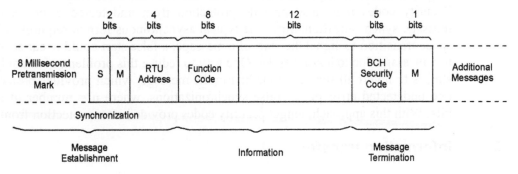

Figure 3.9
Typical asynchronous message format

The message establishment field has three components:
- An 8 millisecond (minimum) pre-transmission mark to condition the modem receiver for the synchronization bits.
- Synchronization: This consists of two bits: a space followed by a mark. The asynchronous interface is designed to start decoding bits after a mark-to-space transition. Therefore the change from the pre-transmission mark to the space provides this transition.

- RTU address: This allows a receiver to select the message addressed to it from the messages to all RTUs on a party line. To avoid any possible mix-ups on addressing the wrong RTU, it is recommended that each RTU in the system has a unique address.

The information field contains 20 bits, of which eight bits are a function code and 12 bits are used for data. For remote-to-master messages, this represents the first message in a sequence, additional messages directly following the first message also transmit information in the RTU address and function code spaces, so that 24 bits of data are transmitted. These 24 bits may contain two 12 bit analog values or 24 device statuses. Additional discussion of the use of the information field is contained in the section 'information transfer'.

The message termination field contains:

- BCH (Bose-Chaudhuri-Hocquenghem) security code, which has five bits and allows the receiving logic to detect most message errors. If an error is detected, the message may then be retransmitted to obtain a correct message.
- End of message mark, which provides the last bit as a mark, so that another message can follow immediately after this message (due to the requirement for a mark-to-space transition for synchronization).

The efficiency of the example format is 12/32 or 37.5% for the first message and 24/32 or 75% for subsequent messages. This is typical for the asynchronous format. The security of the format is provided by the five bit BCH code, which detects all single bit and double bit random errors and all bursts (consecutive bit stream, where first and last bits as a minimum are in error) of five or less. Therefore this BCH code provides good error detection with only a small loss in efficiency. Other equally powerful codes used by several manufacturers are geometric codes wherein parity is used for each word in a message and for correspondingly positioned bits within all words. These codes detect all single, double, and triple bit errors and all bursts of the word length or less (typically 16). Security codes must also provide protection from undetected errors caused by false message synchronization. Since the typical asynchronous format requires only a mark-to-space transition to signal the start of a message, a false start could occur several bits prior to a message due to a noise spike. One way to reduce this problem, commonly called sync slip, is to invert all security code bits in the message, which provides protection equal to one undetected error per 2^n false synchronizations, where n = number of security code bits. With this approach, longer security codes provide better protection from sync slip.

3.4.2   Information transfer

Master-to-remote data transfer

Information transmitted from master to remote is for the purpose of device control, set-point control, or batch data transfer. Due to the possible severe consequences of operating the wrong device or receiving a bad control message, additional security is required for control. This is provided in the form of a sequence of messages, commonly called a select-before-operate sequence, as shown in Figure 3.10

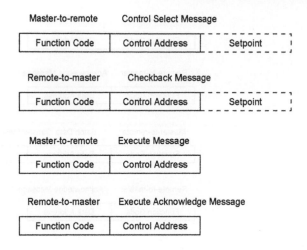

Figure 3.10
Sequence of messages for control

The following explanatory notes apply to Figure 3.10:

- Message establishment and message termination fields are not shown.
- Function code specifies the operation to be performed by the RTU.
- Control address specifies the device or set-point to be controlled.
- Set-point provides the value to be accepted by the RTU.
- A remote-to-master check-back message is derived from the RTU point selection hardware in order to verify that the RTU has acted correctly in interpreting the control selection.

A master-to-remote execute message is transmitted only upon receipt of a proper check-back message.

A remote-to- master execute acknowledge message is a positive indication that the desired control action was initiated.

The above sequence of messages provides additional security by the check-back and execute messages, since undetected errors must occur in the control, select, check-back, and execute messages in order to operate the wrong control device. Before transmission of the above sequence, a control operator or dispatcher performs a similar select-verify-execute-acknowledge sequence via his/her control console.

For certain types of control operations, i.e. raise/lower for electric generating units, the consequences of operating the wrong device are only a single pulse to the wrong unit. Since the automatic generation control system will soon correct this error, there is no serious problem. Therefore, only the first message of the above sequence is transmitted.

Batch data transfers from master to remote include such data as parameters for report by exception and parameters for locally controlled devices. This type of transfer is accomplished by the sequence shown in Figure 3.11.

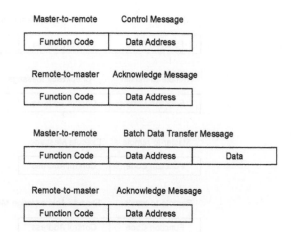

Figure 3.11
Sequence of messages for batch data transfers

The following explanatory notes apply to Figure 3.11:

- Message establishment and message termination fields are not shown.
- Special security precautions are required if a party line communication channel is used, so that other RTUs do not decode a batch data transfer message.

In Figure 3.11 the purpose of the first two messages is to prepare the RTU to receive the longer than normal message. The third message transmits the data and the fourth indicates that the data was correctly received at the RTU.

Remote-to-master data transfer

All remote-to-master data transfer is accomplished with one basic message sequence by using variations in the field definitions to accommodate different types of data. The basic sequence is shown in Figure 3.12

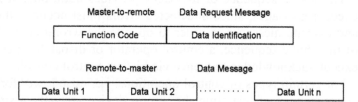

Figure 3.12
Sequence of messages for data acquisition

The following explanatory notes apply to Figure 3.12:

- Message establishment and message termination fields are not shown.
- Function code specifies the type of data to be transferred by the RTU.
- Data identification identifies the amount and type of data requested by the master station.

On each message transmitted by the RTU, (except for messages containing only current data) it is necessary to retain the transmitted message in an RTU buffer so that if the master station does not receive the message correctly, it can request a retransmission. Otherwise this information would be lost.

Three basic types of data are transferred using the sequence of Figure 3.12:

Current data pertaining to the current state of external equipment and processes at the time the data is transmitted to the master station. Data units may be analog values (usually represented by 12 bits per value), binary state of switches (one bit per switch) or the binary state of switches with past changes of state (two bits per switch). Each message may contain many data units of the above types of data, transmitted in predefined order. Messages are usually of fixed length, so that the master station knows the number of data units to expect. If not all data can be transmitted in one message, the master station will request additional messages until all data is received.

Example messages might be:

- 16 analog values (one per data unit)
- 128 status bits (16 per data unit)
- 64 status/status with memory pairs (8 per data unit)
- Combinations of analog values and status bits

Transmitted messages containing memory of past changes of state must be protected by a transmit buffer to avoid any loss of data.

Data snapshot consisting of information stored at the RTU at some previous instant of time (usually commanded by the master station or by a local time source at the RTU). Data units may be analog values (usually 12 bits per value), or memory locations (8 or 16 bits per location).

Since it is often desirable to obtain a simultaneous snapshot from several RTUs, the master station to those RTUs transmits a 'broadcast freeze' command simultaneously. This command has a specific code in place of the RTU address in the message establishment field; all affected RTUs accept this code. The function code indicates the type of data snapshot to be performed.

Data by exception reporting consisting of information concerning the state of external equipment and processes which have changed since the previous reporting. Examples are switches, which have changed state, and analog values, which have changed by more than a preset increment or decrement since the previous report. Since the master station has no way of knowing which values will be reported, each data unit must include, in addition to the new device state or analog value, the point address within the RTU. Also, since the master station does not know the overall message length, these messages must be of fixed length (unused bits may be filled with codes not representing any valid data). Multiple messages may be required to report all changes.

In certain systems it is desirable to record, at the RTU, the time at which each switch changed state. Often termed 'sequence of events', this technique provides information about the operation of field devices by the time order in which state changes occurred. To support this application, each data unit must include, in addition to the new device state and point address, the time to the nearest millisecond.

3.4.3 High level data link control (HDLC) protocol

HDLC has been defined by the international standards organization for use on both multipoint and point-to-point links. Other variations of this protocol include SDLC (synchronous data link control used by IBM) and ADCCP (advanced data communication control procedure used by ANSI). HDLC is a bit-based protocol. Other protocols are based on characters (e.g. ASCII) and are generally slower. It is interesting to note that it is a predecessor to the LAN protocols.

The two most common modes of operation of HDLC are:

- Unbalanced normal response mode (NRM): This is used with only one primary (or master) station initiating all transactions.
- Asynchronous balanced mode (ABM): In this mode each node has equal status and can act as either a secondary or primary node.

3.4.3.1 Frame format

The standard format is indicated in Figure 3.13 below. The three different classes of frames used are:

- Unnumbered frames are used for setting up the link or connection and to define whether unbalanced normal response or asynchronous balanced modes are to be used. There are no sequence numbers contained in these frames; hence they are called unnumbered frames.
- Information frames are used to convey the actual data from the one point to the other.
- Supervisory frames are used for flow control and error control purposes. They indicate whether the secondary station is available to receive the information frames and to acknowledge the frames. There are two forms of error control used – either a selective retransmission procedure because of an error or a request to transmit a number of previous frames.

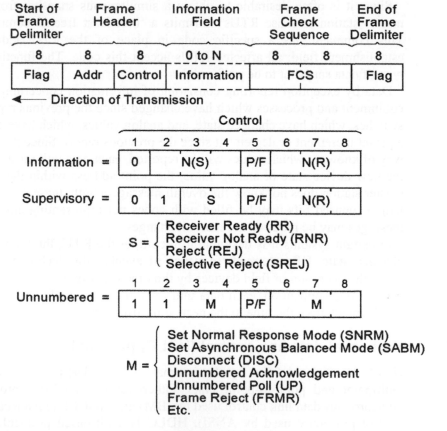

Figure 3.13
HDLC frame format and types

3.4.3.2   Contents of frame

The contents of the frame are briefly as follows:
- The flag character is a byte with the form 01111110. In order to ensure that the receiver always knows that the character it receives is a unique flag character (rather than merely some other character in the sequence), a procedure called zero insertion is followed. This requires the transmitter to insert a '0' after a sequence of five 1's in the text (i.e. non flag characters).
- The frame check sequence (FCS) uses the CRC-CCITT methodology except that 16 ones are added to the tail of the message before the CRC calculation proceeds and the remainder is inverted.
- The address field can contain one of three types for the request or response message to or from the secondary node:
- Standard secondary address
- Group addresses for groups of nodes on the network
- Broadcast addresses for all nodes on the network (here the address contains all 1s)

Where there are a large number of secondaries on the network, the address field can be extended beyond 8 bits by encoding the least significant bit as a 1. This then indicates that there is another byte to follow in the address field.

The control field is indicated in Figure 3.13. Note that the send and receive sequence numbers are important to detect and correct errors in the messages. The P/F bit is the poll/final bit and when set indicates to the receiver that it must respond or acknowledge this frame (again with the P/F bit set to 1).

3.4.3.3   Protocol operation

A typical sequence of operations is given below.
- In a multidrop link, the primary node sends a normal response mode frame with the P/F bit set to 1 together with the address of the secondary.
- The secondary responds with an unnumbered acknowledgment with the P/F bit set to 1. Alternatively if the receiving node is unable to accept the setup command a disconnected mode frame is returned.
- Data is then transferred with the information frames.
- The primary node then sends an unnumbered frame containing a disconnect in the control field.
- The secondary then responds with an unnumbered acknowledgment.

A similar approach is followed for a point to point link using asynchronous balanced mode except that both nodes can initiate the setting up of the link and the transfer of information frames, and the clearing of the point to point link.
- When the secondary transfers the data, it transmits the data as a sequence of information frames with the F bit set to 1 in the final frame of the sequence.
- In NRM mode if the secondary has no further data to transfer, it responds with a receiver not ready frame with the P/F bit set to 1.

3.4.3.4   Error control/flow control

The simplest approach for error control is for a half duplex flow of information frames. Each side of the link maintains a send and receive sequence variable. Whenever the receiving node receives a frame it acknowledges with a supervisory frame with a 'receiver ready' indication together with a receive sequence number acknowledging correct sequence of all frames up to one less than the receive sequence number. When the receiver responds with a negative acknowledgment (REJ) frame with a receive sequence number, the transmitter must transmit all frames from that receive sequence number. This is due to the receiver receiving an out of sequence frame.

It is possible for selective retransmission to be used where the receiver would return a selective rejection frame containing only the sequence number of the missing frame to the transmitter.

A slightly more complex approach is for a point-to-point link using asynchronous balanced mode with full duplex operation where information frames are transmitted in two directions at the same time. The same philosophy is followed as for half duplex operation except that checks for correct sequences of frame numbers must be maintained at both sides of the link.

Flow control operates on the principle that a maximum number of information frames that are awaiting acknowledgment at any time is seven. If seven acknowledgments are outstanding, the transmitting node will suspend transmission until an acknowledgment is received (either in the form of a supervisory frame receiver ready frame or piggybacked in an information frame being returned from the receiver).

If however the sequence numbers at both ends of the link become so out of sequence that the number of frames awaiting acknowledgment exceeds seven, the secondary transmits a frame reject or a command reject frame to the primary. The primary then sets up the link again, and on an acknowledgment from the secondary, both sides reset all the sequence numbers and commence the transfer of information frames.

It is possible for the receiver to run out of buffer space to store messages. In this event it will transmit a 'receiver not ready' supervisory frame to the primary to instruct it to stop sending any more information frames.

3.4.4 The CSMA/CD protocol format

The HDLC protocol describes the complete communications process and provides a complete set of rules for controlling the flow of data around a network. The CSMA/CD protocol is not as comprehensive as HDLC and is concerned with the method used to get data on and off the physical medium. HDLC and CSMA/CD can be incorporated together for a more complete protocol.

The format of a CSMA/CD frame, which is transmitted, is shown in Figure 3.14.

The MAC frame consists of seven bytes of preamble, one byte of the start frame delimiter and a data frame.

The data frame consists of a 48-bit source and destination address, 16 bits of length or type fields, data and a 32-bit CRC field.

The minimum and maximum sizes of the data frames are 64 bytes and 1518 bytes respectively.

Preamble	SFD	Destination Address	Source Address	Length Indicator	Data	Pad (optional)	Frame Check Sequence
7 Bytes	1 Byte	2 or 6 Bytes	2 or 6 Bytes	2 Bytes			4 Bytes

Figure 3.14
Format of a typical CSMA/CD frame

The format of the frame can be briefly described as follows (with reference to each of the fields):

- **Preamble field**
  This allows the receiving electronics of the MAC unit to achieve synchronization with the frame. This field consists of seven bytes each containing the pattern 10101010.
- **Start of frame delimiter (SFD)**
  This contains the pattern 10101011 and indicates the start of a valid frame.
- **Destination and source address**
  Each address may be either 16 or 48 bits. This size must naturally be consistent for all nodes in a particular installation.
- **Date**
  The information to be sent.
- **Length indicator**
  This is a two-byte field, which indicates the number of bytes in the data field.
- **Frame check field**
  This contains a 32-bit cyclic redundancy check that is used for error detection.
  The following sequence is followed for transmission and reception of a frame.
- **Transmission of a frame**
  - Frame contents are first encapsulated by the MAC unit
  - Carrier sense signal is monitored by MAC unit for other transmissions on the media
  - If the media is free, bit stream is transmitted onto the communication medium via the transceiver
  - Transceiver monitors for collisions
  - If a collision, the transceiver turns on the collision detect signal
  - MAC unit then enforces collision by transmitting jam sequence (for LANs, but not necessarily always for radio systems)
  - MAC unit terminates transmission and reschedules a retransmission after a random time interval
- **Reception of a frame**
  - MAC unit detects the presence of an incoming signal from the transceiver
  - The carrier sense signal is switched on to prevent any new transmissions from the MAC unit
  - The incoming preamble is used to achieve synchronization
  - The destination address is checked to see if this is the correct node for reception of this frame
  - Hereafter, validation checks are performed on the frame to confirm that the FCS matches the frame's contents; and it is the correct length

3.4.5 Standards activities

For SCADA systems, the development of standards has been a slow and difficult process. The key existing standard is ANSI/IEEE C37.1-1987, 'design, specification and analysis of systems used for supervisory control, data acquisition and automatic control'. Telecommunication is addressed in several sections of this document as follows:

3	Definitions – provides definitions of many data communication and SCADA terms.
4	Functional Characteristics – provides typical equipment configurations and functional applications.
5.4	Communication – provides interface characteristics, master/remote communication interface, and channel loading calculations.
7.4.2	Communication Security – provides performance requirements for error control.

The C37.1 standard, however, stops short of defining a message standard between master station and RTUs, which is a critical need. This subject was taken up in the soon to be released 'IEEE recommended practice for master/remote communication', which contains the following:

3	Definitions – of terms and concepts.
4	Communication Channels – defines types of channels over which communication may take place.
5	Communication Message Format – provides basic structure and content of messages.
6	Information Field Usage – defines procedures for channel management.
7	Information Field Usage – defines procedures for channel management.

Several vendors have announced plans to develop master/RTU protocols in conformance with this IEEE recommended practice, which is a long overdue step on the way to standardization. Regardless of the protocol used, the recommended practice has many useful guidelines to designing a SCADA communication system.

In the international arena, the ISO (International Standards Organization) reference model for open systems interconnection is preferred. This breaks the communication process up into seven distinct layers, with well-defined interfaces for each layer. The standard of interest is the high level data link control (HDLC) protocol as discussed earlier. This has not been widely used due to security concerns and the low efficiency of the protocol (compared to one designed purely for a SCADA system). But there has been no resolution of these issues insofar as selecting the most appropriate protocol for master to RTU communications is concerned.

Interestingly the LAN standards discussed earlier on (Ethernet, token ring, and token bus) for connecting nodes on the master station are widely accepted and there is little disagreement here. For example, Ethernet (plus TCP/IP) is a widely used LAN standard for master station networks.

3.5 Error detection

Error detection was briefly discussed earlier under protocols. Most error detection schemes involve having redundant bits transmitted with the message to allow the receiver to detect errors in the message bits (and sometimes to reconstruct the message without having to request a retransmission).

3.5.1   Causes of errors

Typically a signal transmitted across any form of transmission medium can be practically affected by four phenomena:
- Attenuation
- Limited bandwidth
- Delay distortion
- Noise

Each of these will be briefly considered.

- **Attenuation**

  As a signal propagates down a transmission medium its amplitude decreases. This is referred to as signal attenuation. A limit should be set on the length of the cable and one or more amplifiers (or repeaters) must be inserted at these set limits to restore the signal to its original level. The attenuation of a signal increases for its higher frequency components. Devices such as equalizers can be employed to equalize the amount of attenuation across a defined band of frequencies.

- **Limited bandwidth**

  Essentially the larger the bandwidth of the medium the closer the received signal will be to the transmitted one. The Nyquist formula to determine the maximum data transfer rate of a transmission line is:

  $$Max\ Transfer\ Rate\ (bps) = 2\ B \log_2 M$$

  where:

  B is the bandwidth in hertz

  M is the number of levels per signaling element.

  For example, with a modem using PSK and four levels per signaling element (i.e. two frequencies) and a bandwidth on the public telephone network of 3000 Hz, the maximum data transfer rate is calculated as:

  $$Maximum\ Data\ Transfer\ Rate = 2 \times 3000 \log_2 4$$
  $$= 12\,000\ bits\ per\ second$$

- **Delay distortion**

  When transmitting a digital signal the various frequency components of the signal arrive at the receiver with varying delays between them. Hence the received signal is distorted with the effects of delay distortion. When the frequency components from different discrete bits interfere with each other, this is known as intersymbol interference.This can lead to an incorrect interpretation of the received signal as the bit rate increases.

- **Noise**

  An important parameter associated with the transmission medium is the concept of signal to noise ratio:

$$Signal\ to\ Noise\ Ratio = 10 \log_{10} \frac{S}{N} \text{ dB}$$

where:

S = the signal noise power in Watts

N = the noise power in Watts

The theoretical maximum data rate of a transmission medium is calculated using the Shannon-Hartley Law, which states:

$$Max\ Data\ Rate = B \log_2 \left(1 + \frac{S}{N}\right) \text{ bps}$$

where:

B = the bandwidth in Hz

S = the signal power in watts

N = the random noise power in watts

For example, with a signal to noise ratio of 100, and a bandwidth of 3000 Hz, the maximum theoretical data rate that can be obtained is:

Maximum information rate = $3000 \log_2 (1 + 100)$

= 19 963 bits per second

There are two approaches to coping with errors in the message; feedback error control in which the receiver detects errors in the message and then requests a retransmission of the message and forward error control where the receiver detects errors in the message and reconstructs the message from the redundant data contained in the message. Forward error control will not be discussed here; refer to the modem section for a full treatment of this subject.

3.5.2 Feedback error control

Message security

It is essential to protect against fake control action and corruption of data resulting from communication noise. Security is achieved by adding a check code to each transmitted message. The concept is for the transmitting station to calculate the check code from the message pattern. The receiving station then repeats the same check code calculation on the message and compares its calculated check code to that of the message received. If they are identical it is assumed that the received message has not been corrupted. If they are different the message is discarded.

Typical security code formats used are:

- **Simple parity check**
  A single bit is added to each byte of the message so that (for example) each group of bits always adds up to an even number.

- **Block check calculation**
  This is an extension on the single parity check in that a new byte is calculated (at the end of the message), based on parity check or a simple arithmetic sum of bits.

- **2-out-of-5 coding**

  Two out of five bits out of each group of five are set at any given time.

- **BCH (Bose-Chaudhuin-Hacquengham)**

  Each block of data (26 bits) is divided by a complex polynomial and the remainder of the division is added to the end of the message block (typically as a 5-bit code).

- **Cyclic redundancy check (CRC-16 or CRC-CCITT)**

  This is similar in concept to the BCH in that the remainder is a 16-bit code, which is appended to the end of the message. The CRC-16 is probably the most reliable security check, which can easily be implemented.

Three of the most commonly used methods for error detection will be discussed in more detail below.

- **Character redundancy checks**

  Before the transmission of the character, the transmitter uses the agreed mechanism of EVEN or ODD parity to calculate the parity bit to append to the character.

  For example:

  If ODD parity has been defined as the mechanism for transmission of ASCII 0100001 this becomes 01000011 to ensure that there are an odd number of 1s in the byte. For an EVEN parity scheme the above character would be represented as 01000010. At the receiving end, parity for the 7 bit data bytes is calculated and compared to the parity bit received. If the two do not agree, an error has occurred.

  Parity error detection is not used much nowadays for communication between different computer and control systems and the more sophisticated algorithms available, such as block redundancy parity check and cyclic redundancy check (CRC) are used.

- **Block redundancy checks**

  Character parity error checking discussed earlier is unacceptably weak in checking for errors. There are two methods of improving on this described below. The parity check on individual characters is supplemented by a parity check on a block of characters.

- **Parity check (vertical/longitudinal redundancy check)**

  In the block check strategy, message characters are treated as a two dimensional array. A parity bit is appended to each character. After a defined number of characters, a block check character (BCC), which represents a parity check of the columns, is transmitted. Although column parity (also referred to as vertical redundancy check) is better than the character parity error checking, it still cannot detect an even number of errors in the rows.

TRANSMITTED	RECEIVED	
Message	Double-bit error in 1 row	Double-bit error in 2 rows
c 11000011	11000011	11000011
f 01100110	01100110	01100110
y 11111001	11**00**001	111**00**001
u 01010101	01010101	010**01**101
BCC 00001001	00001001	00001001
00001001 BCC calculated by receiver	00010001 (error detected)	00001001 (error not detected)

Table 3.1
Vertical/longitudinal redundancy check using EVEN parity

- **Arithmetic checksum**

  An extension of the vertical redundancy check is to use an arithmetic checksum, which is a simple sum of characters in the block. This provides even better error-checking capabilities and also increases the overhead as two bytes now have to be transmitted.

TRANSMITTED	RECEIVED		
Message	Double-Bit error in 1 row	Double-bit error in 2 rows	Single-bit Error in 2 columns
c 1000011	1000011	1000011	100001**0**
f 1100110	1100110	1100110	110011**1**
y 1111001	11**00**001	11**00**001	1111001
u 1010101	1010101	10**01**101	1010101
BCC 101110111	101110111	101110111	101110111
BCC calculated by receiver	101011111 (detected)	101010111 (detected)	101110111 (not detected)

Table 3.2
Block redundancy – checksum

- **Cyclic redundancy check (CRC)**

  This provides a worse case probability of detecting errors of 99.9969%.

There are two types of CRC calculations performed.

- CRC-CCITT (popular in commercial systems)
- CRC-16 (popular in industrial systems)

The CRC checksum is calculated by dividing the message by a defined number (known by the receiver and transmitter of the message) and calculating the remainder. The remainder is known as the CRC checksum and is appended onto the end of the message.

CRC example

The following equation can be proven:

$$\frac{Message \times 2^{16}}{Divisor} = Quotient + Remainder$$

where:

- Message is a stream of bits, for example; the ASCII sequence of H E L P with even parity.

[01001000]	[**1**1000101]	[**1**1001100]	[01010000]
H	E	L	P

- 2^{16} effectively (in multiplying) add on 16 zeros to the right hand part of the message.
- Divisor is a number, which is divided into the message $\times 2^{16}$ number.
- Quotient is the result of the division and is not used.
- The remainder is the value left over from the result of the division and is the CRC checksum (a two-byte number).

3.6 Distributed network protocol

3.6.1 Introduction

The distributed network protocol is a data acquisition protocol used mostly in the electrical and utility industries. It is designed as an open, interoperable and simple protocol specifically for SCADA controls systems. It uses the master/slave polling method to send and receive information, but also employs sub-masters within the same system. The physical layer is generally designed around RS-232 (V.24), but it also supports other physical standards such as RS-422, RS-485 and even fiber optic. There is large support within the SCADA industry to use DNP as the universal *de facto* standard for data acquisition and control.

3.6.2 Interoperability

The distributed network protocol is an interoperable protocol designed specifically for the electric utilities, oil, gas, and water/waste water and security industries. As a data acquisition protocol, the need to interface with many vendors equipment was and is necessary. By having a certification process, the protocol ensures that different manufacturers are able to build equipment to the DNP standard. This protects the end user when purchasing a certified DNP device. As more and more manufacturers produce DNP certified equipment, the choices and confidence of users will increase.

3.6.3 Open standard

The DNP was created with the philosophy of being a completely open standard. Since no one company owns the DNP standard it means that producers of equipment feel that they have a level playing field on which to compete. This allows different manufactures to have equal input into changes to the protocol. In addition, it means that the cost to develop a system is reduced. The producer does not need to design all parts of the SCADA system. In a proprietary system, the manufacturer usually has to design and produce all parts of the SCADA system, although some of those parts may not be so profitable. One manufacturer is free then to specialize on a few products that are its core business.

3.6.4   IEC and IEEE

DNP is based on the standards of the International Electrotechnical Commission (IEC) Technical Committee 57, Working Group 03 who have been working on an OSI 3 layer 'Enhanced Performance Architecture' (EPA) protocol standard for telecontrol applications. DNP has been designed to be as close to compliant as possible to the standards as they existed at time of development with the addition if functionality not identified in Europe but needed for current and future North American applications. Recently DNP 3.0 was selected as a recommended practice by the IEEE C.2 task force, remote terminal unit to intelligent end device's communications protocol.

3.6.5   SCADA

The DNP is well developed as a device protocol within a complete SCADA system. It is designed as a data acquisition protocol with smart devices in mind. These devices then can be coupled as a multi-drop fieldbus system. The fieldbus DNP devices are integrated into a software package to become a SCADA system. DNP does not specify a single physical layer for the serial bus (multi-mode) topology. Devices can be connected by 422 (four wire), 485 (two wire), modem (Bell 202) or with fiber optic cable. The application program can integrate DNP with other protocols if the SCADA software permits. Using tunneling or encapsulation the DNP could be connected to an intranet or the Internet.

3.6.6   Development

The specification was first developed by the GE Harris Company but has been released under the DNP User Group since 1992. Now over 100 vendors offer DNP V3.0 products. These products range from master stations to intelligent end devices. The protocol is designed so that a manufacturer can develop a product that supports some but not all of the functions and services that DNP supports. The DNP 3.0 was derived from an earlier version of the IEC 870.5 specs. The DNP users group now controls the documenting and updating of the protocol. For users of the DNP a copy of the protocol can be purchased through http://www.dnp.org

3.6.7   Physical layer

The physical layer of DNP is a serial bit oriented asynchronous system using 8 data bits, 1 start bit, 1 stop bit and no parity. Synchronous or asynchronous is also allowed. It has two physical modes of operation, direct mode (point-to-point) or serial bus mode (multi-drop). The two modes are not usable at the same time. Both modes can be half or full duplex. With either mode, a carrier detection system must be used. The DNP protocol is a modified master/slave system. Multi-masters are allowed but only one device can be a master at a time. There are possible collisions on the system. The configuration of the physical layer determines the method of collision avoidance or recovery. The DNP can prioritize devices in a multi-master mode.

3.6.8   Physical topologies

The DNP protocol supports five communication modes, two-wire point-to-point, two wire multidrop, four-wire point-to-point, four-wire multidrop, and dial up modems. A system with only two nodes, a master and a slave is called a direct bus. If the system is multidrop with multiple nodes, it is called a serial bus. Both of these systems can use two or four wire connection methods. The two wire method can only run half duplex while the four wire method can run either half or full duplex. The DNP supports multiple master, multiple slave and peer-to-peer communications.

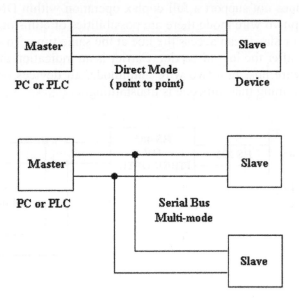

Figure 3.15
Direct and serial modes

3.6.9   Modes

Two wire point-to-point

The DNP protocols physical layer supports point-to-point communications. The two wire half-duplex mode usually uses RS-485 or a two-wire modem as a physical system. If a modem is used then the interface to the modem usually uses the V.24 ITU standard (RS-232). The two wire mode does not support a full duplex operation within DNP, only half-duplex. With the point-to-point mode there is no possibility of collisions. The master transmits the frame and then the slave responds. The only concern is the time it takes for a response due to propagation delays. There is a configuration in DNP for setting up this time.

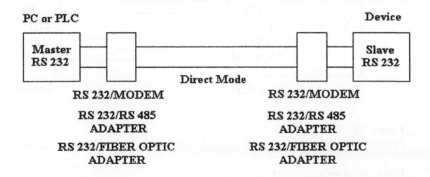

Figure 3.16
Two wire direct mode

Two wire multidrop

The DNP physical layer supports multi-point communications. The two wire multidrop mode usually uses RS-485, fiber optic or Bell 202 modems as a physical system. The two wire mode does not support a full duplex operation within DNP, only half-duplex. With the multidrop two wire mode there are possibilities of collisions. This is possible because two masters or slaves can access the line at the same time. To overcome this DNP inserts a time delay after the loss of carrier. Carrier is an indication that someone is transmitting on the two wire bus. In a two wire multimode, all devices on the line must have some way of determining that someone is transmitting.

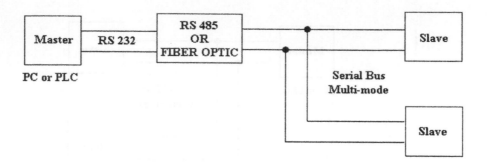

Figure 3.17
Two wire multimode

Four wire point-to-point

Four-wire point-to-point is used within DNP as a full duplex master to slave system. The physical standards used are RS-422, RS-232 and four wire modems. Since this mode is only point-to-point, there is no problem with collisions. However, V.24 is used as a handshaking system to control the communications. This includes DCD (data carrier detect). Propagation delays are also used to allow the devices time to detect the loss of carrier. Four wire communications allows for true full duplex communication, but in practice, it is rarely implemented.

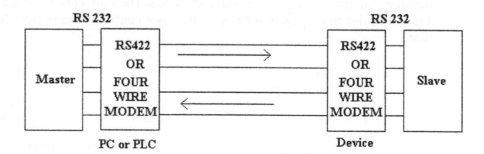

Figure 3.18
Four wire direct mode

Four wire multi-point

DNP allows for a four wire multi-point mode. This mode can use half or full duplex communications, although again full duplex is rarely used. The reason full duplex is not used is because of the complexity of collision avoidance. Having multiple devices all talking at the same time in both directions is difficult at best to implement. The master does not have a problem with collisions but may have problems with other primary masters accessing the line before the slave has a chance to respond. The slave can have many problems with collisions because many of them may want to respond to their masters at the same time. One way that DNP uses to handle these collisions is to allow the slaves to collide. This causes the second master to time out and the primary master to gain access to the bus. The master can then send out high priority messages.

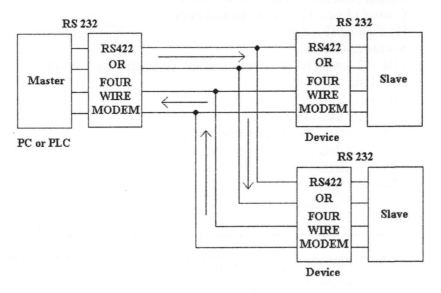

Figure 3.19
Four wire serial mode

Dial up modem

The DNP supports the use of a dial up modem mode. This mode is a point-to-point circuit. It usually uses V.24 as a connection system (RS-232). DCD is used differently in this case because carrier detect in modems means that a line has been established, not that data is being sent. The RTS line is placed high to tell the modem that the DTE wishes to send data. The CTS is placed high by the modem to tell the DTE it is OK to send data. There is no way for the local end to tell the remote end that data is being transferred. It is then up to the remote end to be able to detect data coming in.

Figure 3.20
Dial up modem mode

3.6.10   Datalink layer

The datalink layer of the DNP defines the frame size, shape, length and contents. DNP uses the convention of the octet instead of byte. The DNP uses hexadecimal as a language within the frame. The frame is laid out as follows

3.6.10.1   Frame outline

Start (2 octets) 0X0564   (0000010100100000)
Length (1 octet) 5 to 255   (DECIMAL)
Control (1 octet) includes function code
Destination (2 octets)
Source (2 octets)
CRC (2 octets) for the length, control, destination and source
User Data (16 octets)
CRC (2 octets) for the user data above
More user data (16 octets)
CRC (2 octets) for the user data above only
More user data (1 to 16 octets) variable
CRC (2 octets)
END

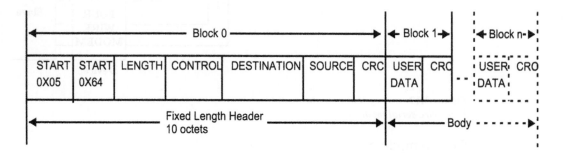

Figure 3.21
DNP packet format

Notice that the user data can go on until the maximum number of octets is reached. This is determined by the length octet. The maximum octets of data is 255 and the minimum is 5. The last user data may have less than 16 octets. Each CRC is calculated for that user data, not for the whole frame.

3.6.10.2   Function codes

There are four bits used in the control octet to determine how the data link will handle the frame. There are six basic function codes:

- **Reset**
  This function code is used to synchronize a primary and secondary station for further send-confirm transactions.

- **Reset of user process**
  This function code is used to reset the data link user process

- **Test**
  The Test command is used to test the state of the secondary data link

- **User data**

  The 'user data' function is used to send confirmed data to a secondary station.

- **Unconfirmed user data**

  This function is used to send user data to the secondary station without needing confirmation.

- **Request link status**

  This command is used to request the status of the secondary data link.

3.6.10.3 Examples of transmission procedures

Reset of secondary link

In the figure below, a primary station sends a send-confirm-reset frame to a secondary station. The secondary station receives the message and responds with an ACK confirm frame.

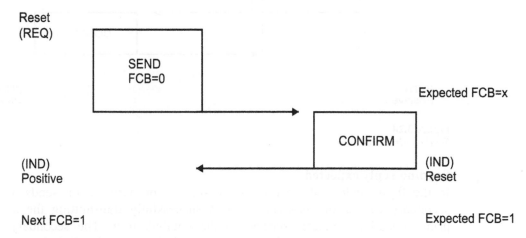

Figure 3.22
Reset of secondary link

Reset of user process

In the figure below, a primary station sends a send-confirm-reset user process frame to a secondary station. The secondary station receives the message and responds with an ACK confirm frame.

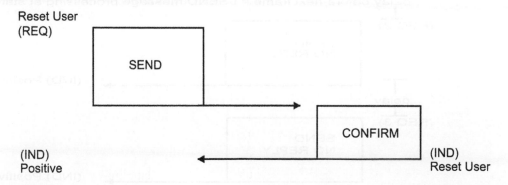

Figure 3.23
Reset of user process

Send/confirm user data

In the figure below, the designated master station acting as a primary station sends a send-confirm frame to a non-master station acting as a secondary station. This is the first frame with PCV valid after the link was reset so FCB = 1 in the send frame. The secondary station expects FCB to be 1 since this is the first frame after the link was reset and sends a confirm frame. The master station upon receiving the confirm assumes the message was correctly received and indicates success to the master station data link user.

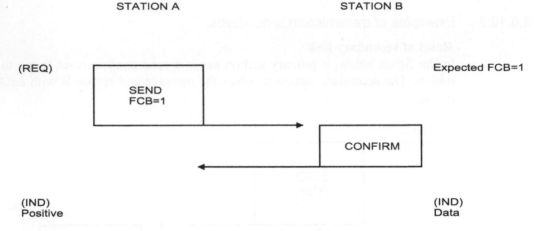

Figure 3.24
Send/confirm user data

Send/no reply expected

In the figure below, the master or non-master primary station sends 3 frames to the secondary master or non-master. Upon successfully transmitting the send frame, the primary station indicates success to the datalink user. The secondary station, upon reception of a valid frame indicates data availability to the data link user.

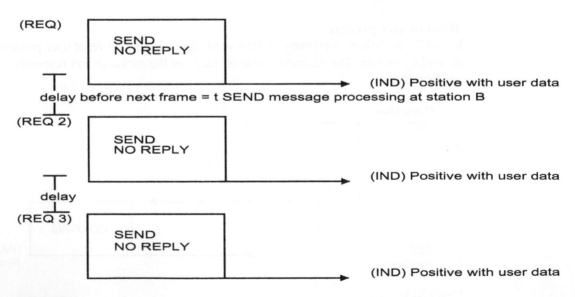

Figure 3.25
Send/no reply expected

Send/NACK

In the figure below, a non-master primary station sends a frame to the master secondary. Upon reception of the first confirm, the primary indicates success to the data link user. The primary sends a second frame to the secondary. The secondary master decides that it cannot accept any frames at this time and sends a NACK frame back. The primary, after receiving this NACK, will fail the transaction and send a negative indication to the data link user.

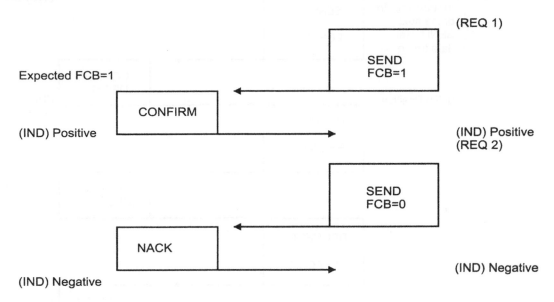

Figure 3.26
Send/NACK

Request/respond

In the Figure 3.27, a primary station sends consecutive frames to a secondary station. When the secondary station cannot receive any more frames, the confirm message contains the DFC bit set. The primary station will, upon reception of the confirm, stop sending data frames to the secondary station but will instead periodically request the status of the secondary by sending a request-respond frame. The secondary will respond to the request frame with the current state of the DFC. If the secondary is ready to receive more data, the DFC returned will be 0 otherwise the DFC returned will be 1. When the primary station recognizes DFC = 0 in the respond frame, the transmission of send frames will continue.

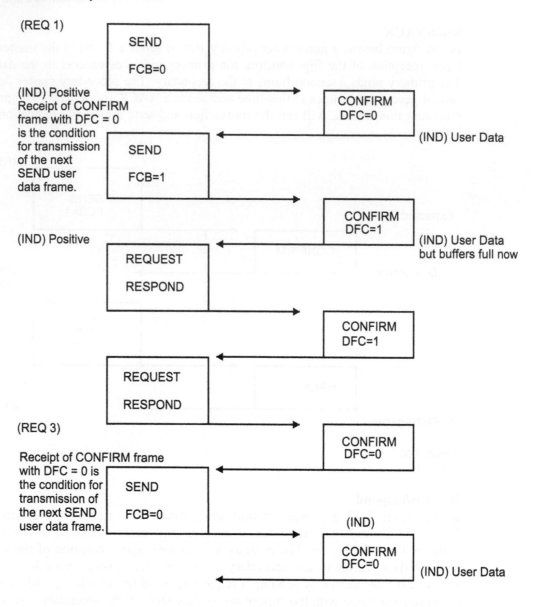

Figure 3.27
Request/respond

3.6.11 Transport layer (pseudo-transport)

The distributed network protocol does not support a true transport layer as defined by the ISO open system interconnection model. It does support a pseudo-transport layer known as the super-data link transport protocol. This is because some of the functions of the data link layer do not strictly meet the ISO OSI model. These functions are then moved out of the data link layer and placed in this pseudo-transport layer. These data link functions consist of breaking the transport service data unit (TSDU) into smaller sequenced frames called link service data units (LSDU). Each of these frames has transport protocol control information. The maximum size of an LSDU is 249 octets. This is done to reduce the length of packets in case of errors. If a packet is in error, then a retry will be initiated. A shorter packet means that the retries will be quicker.

3.6.12   Application layer

The DNP supports an application layer by defining an extensive data object library, function codes and message formats for both the requestor and the response devices. These are used in the USER layer to build a final application. Once this application is built and the data objects, function codes and message formats are absorbed the application becomes the application layer. A complete list of data objects and function codes can be found in the DNP version 3.0 standard document available from DNP users group http://www.dnp.org. And other information can be found at Harris controls division. http://www.harris.com/harris/search.html

3.6.13   Conclusion

The distributed network protocol only supports the physical layer, data link layer and application layer within the open system interconnection model. The physical layer is the least supported. DNP is based on the enhanced protocol architecture (EPA), a protocol standard for telecontrol applications. It supports advanced RTU functions and messages larger than the normal frame length. It takes user data and breaks it up into several sequenced transport protocol data unit (TPDU) each with transport protocol control information (TPCI). The transport protocol data unit is sent to the data link layer as a link service data unit. The receiver receives multiple sequenced transport protocol data unit (TPDUs) from the data link layer and assembles them into one transport service data unit (TSDU).

There is no official compliance testing, but there is help online. If a vendor claims to comply with one of the DNP v3.00 subset definitions, then the device is definitely interoperable. Of course interoperable does not mean efficient. It is often best to stick with one supplier when possible. The distributed network protocol is truly an open non-proprietary interoperable protocol.

3.7   New technologies in SCADA systems

A few of the new developments that are occurring in SCADA technology will be briefly listed below. The rapid advance in communications technology is an important driving force in the new SCADA system.

3.7.1   Rapid improvement in LAN technology for master stations

LANs are increasingly forming a key component of the master stations with dual redundant LANs being able to provide very reliable systems. The movement to higher speed LANs (100 Mbit/sec up from 10 Mbit/sec) are providing faster response times.

3.7.2   Man machine interface

Typical areas where improvements are occurring are:
- Improved graphics on the VDUs with the operator planning and zooming on the system on-line to arrive at any given subset of the network.
- Improved response times on the operator interfaces.

3.7.3 Remote terminal units

- Decentralized processing of the data at the RTU rather than the master station
- Further decentralized gathering of data from intelligent instruments, which transfer the data back to the RTU over a communication network
- Redundancy of RTUs is easily implemented on I/O, CPU, power supplies etc
- Multiple communications with multiple masters with partitioned (separate) databases for each master station
- User generated programs could be run in the RTU to reduce the number of alarm traffic to master station (by combining alarms, filtering or irrelevant alarms)
- Checking on the validity of real-time data received
- Inter RTU communications (rather than through the master stations)
- Sophisticated man-machine interfaces directly connected to RTU

3.7.4 Communications

- Open standards (i.e. non-vendor specific) are appearing to interface RTUs to the master stations
- Spread spectrum satellite – an improved, low cost and low power method of transferring data over a satellite system for remote site RTUs
- Fiber optics – lower cost and ease of installation is making this an attractive option
- Meteor trail ionization – this is becoming an effective technology today especially where it is difficult to justify the cost of a satellite system

3.8 The twelve golden rules

A few rules in specifying and implementing a SCADA system are listed below:

- Apply the 'KISS' principle and ensure that the implementation of the SCADA system is simple.
- Ensure that the response times of the total system (including the future expansion) are within the correct levels (typically less than one-second operator response time).
- Evaluate redundancy requirements carefully and assess the impact of failure of any component of the system on the total system.
- Apply the open systems approach to hardware selected and protocols communication standards implemented. Confirm that these are indeed TRUE open standards.
- Ensure that the whole system including the individual components provide a scaleable architecture (which can expand with increasing system requirements).
- Assess the total system from the point of view of the maximum traffic loading on the RTU, communication links and master stations and the subsequent impact on hardware, firmware and software subsystems.
- Ensure that the functional specification for the system is clearly defined as far as number of points are concerned, response rates and functionality required of the system.

- Perform a thorough testing of the system and confirm accuracy of all data transferred back, control actions and failure of individual components of system and recovery from failures.
- Confirm operators of individual components of the system in the (industrial) environment to which they would be exposed (including grounding and isolation of the system).
- Ensure that all configuration and testing activities are well documented.
- Ensure that the operational staffs are involved with the configuration and implementation of the system and they receive thorough training on the system.
- Finally, although the temptation is there with a sophisticated system, do not overwhelm the operator with alarm and operational data and crowded operator screens. Keep the information of loading to the operator clear, concise and simple.

4

Landlines

4.1 Introduction

Landline is the term used to describe a communications link that is constructed of one or more types of cable. A landline may be a privately owned cable in a factory or on an industrial site, installed, used and owned by a company. Alternatively, it may be a connection through the public network of a telephone company.

The choice of landline to be used depends on the application, the costs involved, the data speeds required, the frequency of access required to the RTU and the available communication services and technologies at this site.

This chapter will examine the different options available and will put them into perspective so a comparison can be made between the options and then a logical decision made as to which option is best for the application. First an examination will be made of noise and interference on cables, how to reduce this noise and then at the different types of cables available.

The user may be considering the installation of new cables, or the use of existing cables. In either case, a good understanding of cable technology is required.

4.2 Background to cables

The connecting data cable is defined as the complete assembly required to join two ends together, including the conductors (or fibers), the shielding, the insulation, the supports, the connectors and the terminations at both ends. The main factors that need to be considered when specifying data communication cables are:

- The type of application
- The type of cable and its shielding (affects noise performance)
- The intended installation method and the cable route (affects noise performance and determines the type of cable that may be used)
- The cable resistance (affects the signal attenuation)
- The cable capacitance (affects the frequency bandwidth and data transfer rates)
- Connections and terminations (affects installation cost and time, and signal attenuation).

In contrast to power cables, the sources of external electrical noise and the cable's ability to exclude it are important aspects of data cable design and installation.

A useful reference on the subject of noise is the IEEE 518-1982 guide entitled 'IEEE Guide for the Installation of Electrical Equipment to Minimize Electrical Noise Inputs to Controllers from External Sources.'

4.3 Definition of interference and noise on cables

Interference and noise are important factors to consider when designing and installing a data communication system, with particular consideration required to avoid electrical interference. Noise can be defined as the random generated undesired signal that corrupts (or interferes with) the original (or desired) signal.

Noise may be generated in the system itself (internal noise), or from an outside source (external noise). Some typical examples of these sources are outlined below –

Internal noise

- Thermal noise – due to electron movement through the circuit
- Imperfections in circuit design
- Stray signals from oscillators and amplifiers
- Intermodulation of stray low level RF produced by internal circuits

External noise

- Natural origins – electrostatic interference, electrical storms
- Electromagnetic interference (EMI)
- Radio frequency interference (RFI)
- Crosstalk

The important characteristic to recognize with noise is that it is random in nature and therefore the interference it produces is unpredictable. This unpredictability makes the design of the cable communications systems quite challenging.

Noise itself is only important if it is measured in relation to the communication signal, which carries the data information. As discussed earlier, the measure of performance of a communications link at any point is given by the signal to noise ratio. This same principle applies to cables.

The signal to noise (SNR) ratio is expressed (in dB) as:
SNR = 10 log (S/N) dB

In a cabling system, an SNR of 20 dB is considered noisy while SNR of 60 dB is considered acceptable. The opinion as to whether a SNR is good or bad will usually be determined by the required data rate over the cable in question. The higher the SNR, the easier it is to provide acceptable performance, with simpler circuitry and cheaper cabling.

In data communications, a more relevant measure of the performance of the communications link is the bit error rate (BER), which is a measure of the number of successful bits received compared to bits that are in error. A BER of 10^{-6} (one errored bit in a million) is considered only fair in performance on a bulk data communication system with high data rates. A BER of 10^{-12} (one error bit in a million) is considered excellent. On industrial systems with low data requirements, a BER of 10^{-4} could be quite acceptable because time varying data (which varies gradually) is transferred, rather than exact accountable data (e.g. EFTPOS).

There is a relationship between SNR and BER. As the SNR increases, the error rate drops off rapidly as shown on the graph below. Most communication systems start to provide reasonably good BERs when the SNR is above 20 dB.

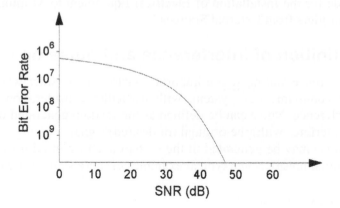

Figure 4.1
Example of a bit error rate versus signal to noise ratio curve

The shape of the BER versus SNR curve will depend significantly on the type of communications link and the type and quality of equipment connected at each end.

4.4 Sources of interference and noise on cables

Noise is normally introduced into cable circuits through electrostatic (capacitive) coupling, magnetic (inductive) coupling and resistive coupling. The reduction of these noise signals takes the form of shielding and twisting of signal leads, proper grounding, separation, and good insulation.

Shielding is the protection of the signal wires from noise or unwanted signals.

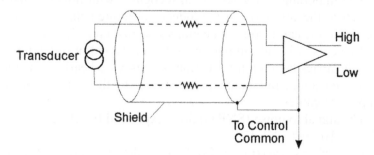

Figure 4.2
Typical shield

The purpose of the shield is to reduce the magnitude of the noise coupled into the low-level signal circuits by electrostatic or magnetic coupling. The shield may be considered an envelope that surrounds a circuit to protect the cable from the coupling. This is illustrated in Figure 4.2.

4.4.1   Electrostatic coupling

Electrostatic or capacitive coupling of external noise is illustrated in Figure 4.3. The external noise source couples the noise into the signal wires through capacitors C_1 and C_2 and the resulting flow of current produces an error voltage signal across R_1, R_2, (the cable resistance) and R_L. The error signal is proportional to the length of the cable leads, the resistance of the cable leads, the amplitude and the frequency of the noise signal and the relative distance of the cable leads from the noise source.

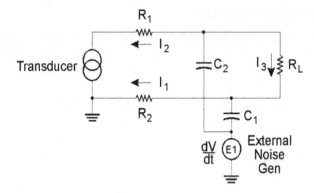

Figure 4.3
Electrostatic coupled noise

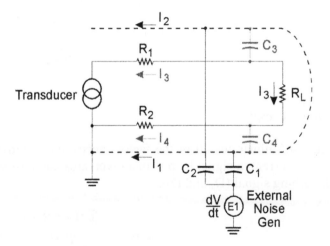

Figure 4.4
Use of shield to reduce electrostatic noise

The noise due to electrostatic coupling can be reduced by the use of shielded wire, by separation and by twisting of the leads. As the separation between the noise source and the signal wires is increased, the noise coupling is thereby reduced. Twisting of the leads provides a balanced capacitive coupling which tends to make $C_1 = C_2$. Therefore, the induced voltages at the load will be equal in magnitude but opposite in direction and should cancel each other out.

The use of a shield to reduce electrostatic noise is illustrated in Figure 4.4. The noise-induced currents now flow through the shield and return to ground instead of flowing through the signal wires. With the shield and signal wire tied to ground at one end, a zero-potential difference would exist between the wires and the shield. Hence, no signal current flows between wire and shield.

The quality of the shield will determine how large C_1 and C_2 are compared to C_3 and C_4. The better the quality shield (possibly 2 or 3 shields on one cable), the higher the value of C_1 and C_2, and the lower the value of C_3 and C_4. The lower the value of C_3 and C_4 the less the noise value induced into the signal cables.

4.4.2   Magnetic coupling

Magnetic coupling is the electrical property that exists between two or more conductors, such that when there is a current change in one, there will be a resultant induced voltage in the other conductor. Figure 4.5 illustrates a disturbing wire (noise source) magnetically coupling a voltage into the signal circuit.

The alternating magnetic flux from the disturbing wire, induces a voltage in the signal loop, which is proportional to the frequency of the disturbing current, the magnitude of the disturbing current and the area enclosed by the signal loop and is inversely proportional to the square of the distance from the disturbing wire to the signal circuit.

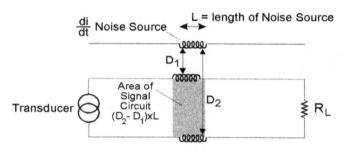

Figure 4.5
Magnetic noise coupling

Figure 4.5 illustrates all of the factors necessary to introduce an error voltage rate of change of current, a signal loop with a given area and a separation of the conductors from the disturbing signal (D_1 and D_2).

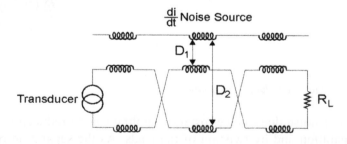

Figure 4.6
Reducing magnetic noise by twisting of wires

A common method of reducing the effect of magnetic coupling is the use of twisted conductors in the signal circuits, as illustrated in Figure 4.6. The distance of these two signal wires with respect to the disturbing wire is approximately equal and the area of the circuit loop is almost zero. Reducing this area to practically zero will reduce the voltage induced by the magnetic field to almost zero due to the equal magnitude of current induced in each lead that will result in a near zero net circulating current. (The currents will induce voltages in the load that are equal and opposite in magnitude and will therefore cancel.)

Employing a shield made of a high ferrous content material around the signal wires can also reduce magnetic coupling. This shield is effective because the magnetic field produces eddy currents in the shield, which will produce magnetic flux in the opposite direction to the inducing flux and will oppose the original magnetic field. A sketch of this type of noise reduction is illustrated in Figure 4.7. This type of shield is very rare and would have to be specially manufactured upon request. The use of high ferrous content conduits is sometimes used but these are subject to corrosion problems.

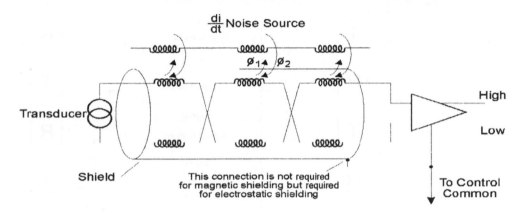

Figure 4.7
Effects of shield in reducing magnetic coupling

4.4.3   Impedance coupling

Impedance coupling (as illustrated in Figure 4.8) is the electrical property that exists when two or more signal wires share the same common return signal wire. If there is any resistance in the common return wire then the signal current from any one of the loads will cause the voltage to rise at all the loads. In addition, noise that induces current flow into the common return will cause noise voltage at all the loads.

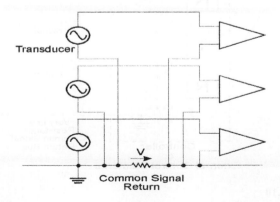

Figure 4.8
Resistance or impedance coupling

To avoid impedance coupling in signal circuits, either of the following can be carried out:

- Employ a low-resistance wire or bus for the common return when a common return cannot be avoided. (For critical applications both the resistance and the inductance of the bus should be minimized.)
- Wherever possible, employ separate signal return leads.

A few alternative solutions to the problem of impedance coupling are indicated in the following diagrams. Figure 4.9 indicates the ideal approach of separate signal returns. Here the common return conductor is reduced to a single terminal point on one side of the links. Figure 4.10 illustrates the use of a large low impedance return bus. Note that the individual returns from each transmitter and receiver should also be as low impedance as possible. Figure 4.11 illustrates a compromise approach where there are too many transmitters and receivers to justify each one getting a single pair of cables.

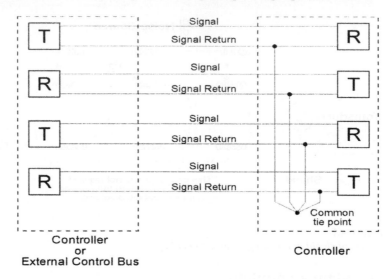

Figure 4.9
Cabling system illustrating individual signal returns

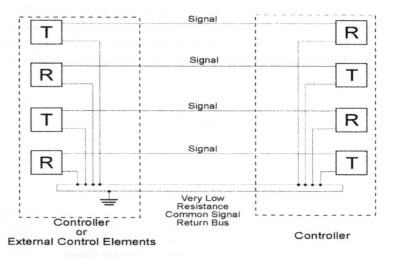

Figure 4.10
Cabling system illustrating common signal returns

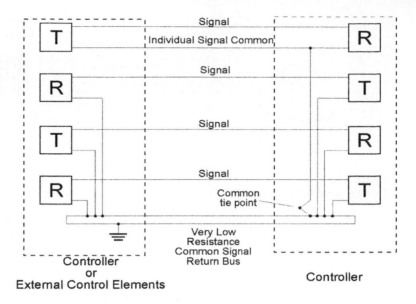

Figure 4.11
Cabling for combined individual and common system returns

4.5 Practical methods of reducing noise and interference on cables

A few practical methods of reducing noise will be examined in this section. These include a quantitative consideration of the effects of shielding, cable spacing, earthing (or grounding) and a note of specific problem areas on which to concentrate.

4.5.1 Shielding and twisting wires

It is theoretically possible to almost eliminate both electric and magnetic field noise (and thus the need for a shield) by twisting the signal cable pairs. Magnetic interference reduction can vary by a factor of 14 for 4-inch lay (or 3 twists per foot) to 141 for 1-inch lay (or 12 twists per foot).

Electrostatic coupling can be reduced by a factor of 103 for a copper braid shield (with 85% coverage) to a factor of 6610 for an aluminum Mylar tape shield. These shielded wires are normally effective for magnetic coupling; hence twisting of the pairs is also desirable within the shield.

It is important that the shield is earthed (or grounded) at one point only, so that all possible ground loops are avoided. This means that the shield envelope should have an insulated jacket to prevent the possibility of the shield earthing and creating multiple grounds. If there is a requirement to earth the shield at both ends, ground loops can be eliminated by decoupling the input amplifier (as in Figure 4.12) or using optically isolated signals (Figure 4.13).

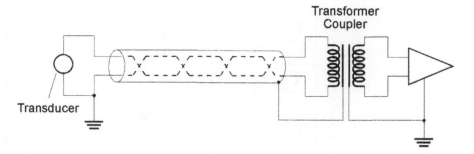

Figure 4.12
Transformer-coupled input

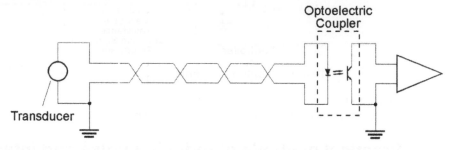

Figure 4.13
Optoelectronic coupler circuit

4.5.2   Cable spacing

In the practical world where there are normally many different cabling systems at one site, a system has been developed to classify all wiring in a certain class of susceptibility to interference and to group the classes in an orderly manner as indicated below.

4.5.2.1   Wiring levels and classes

There are four basic levels or classes of wiring, which can be identified. These classes or levels are listed below with typical signals defined for each level.

Level 1 – high susceptibility

- Analog signals of less than 50 V and digital signals of less than 15 V
- Common returns to high-susceptibility equipment
- Control common tie (CCT)
- DC power supply buses feeding sensitive analog hardware
- All wiring connected to components associated with sensitive analog hardware (e.g. strain gauges, thermocouples, etc)
- Operational amplifier signals
- Power amplifier signals
- Output of isolation amplifiers feeding sensitive analog hardware
- Telephone circuits
- Logic buses feeding sensitive digital hardware
- All signal wires associated with digital hardware

Level 2 – medium susceptibility

- Analog signals greater than 50 V and switching circuits
- Common returns to medium-susceptibility equipment

- DC bus feeding digital relays, lights and input buffers
- All wiring connected to input signal conditioning buffers
- Lights and relays operated by less than 50 V
- Analog tachometer signals

Level 3 – low susceptibility

- Switching signals greater than 50 V, analog signals greater than 50 V, regulating signals of 50 V with currents less than 20 A, and AC feeders less than 20 A.
- Fused control bus 50–250 V DC
- Indicating lights greater than 50 V
- 50–250 V DC relay and contactor coils
- Circuit breaker coils of less than 20 A
- Machine fields of less than 20 A
- Static master reference power source
- Machine armature voltage feedback
- Machine ground-detection circuits
- Line-shunt signals for induction
- All AC feeders of less than 20 A
- Convenience outlets, rear panel lighting
- Recording meter chart drives
- Thyristor field exciter AC power input and DC output of less than 20 A

Level 4 – power

- AC and DC buses of 0–1000 V with currents of 20–800 A
- Motor armature circuits
- Generator armature circuits
- Thyristor AC power input and DC outputs
- Primaries and secondaries of transformers above 5 kVA
- Thyristor field exciter AC power input and DC output with currents greater than 20 A
- Static exciter (regulated and unregulated) AC power input and DC output
- 250 V shop bus
- Machine fields over 20 A

4.5.2.2 Class codes

Within a level, conditions may exist that require specific high quality cables to be used. However, the use of higher quality cabling does not allow regrouping. A class coding system similar to the following may identify the conditions in each level –

A analog inputs, outputs
B pulse inputs
C contact and interrupt inputs
D decimal switch inputs
E output data lines
F display outputs, contact outputs
G logic input buffers

S special handling of special levels may require special spacing of conduits and trays, such as signals from commutating field and line resistors, or signals from line shunts to regulators, or power >1000 V or >800 A, or both

U high voltage potential unfused greater than 600 V DC

4.5.3   Tray spacing

The tables are given below. Table 4.1 indicates the minimum distance in inches between the top of one tray and the bottom of the tray above, or between the sides of adjacent trays.

LEVEL	1	2	3	3S	4	4S
1	0	1	6	6	26	26
2	1	0	6	6	18	26
3	6	6	0	0	18	12
3S	6	6	0	0	8	18
4	26	18	18	8	0	0
4S	26	26	12	18	0	0

Table 4.1
Tray spacing (inches)

LEVEL	1	2	3	3S	4	4S
1	0	1	4	4	18	18
2	1	0	4	4	12	18
3	4	4	0	0	0	8
3S	4	4	0	0	6	12
4	18	12	0	6	0	0
4S	18	18	8	12	0	0

Table 4.2
Tray–conduit spacing (inches)

LEVEL	1	2	3	3S	4	4S
1	0	1	3	3	12	12
2	1	0	3	3	9	12
3	3	3	0	0	0	6
3S	3	3	0	0	6	9
4	12	9	0	6	0	0
4S	12	12	6	9	0	0

Table 4.3
Conduit spacing (inches)

4.5.4   Earthing and grounding requirements

The earth (or ground) is defined as a common reference point for all signals in equipment situated at zero potential. Below 10 MHz, the principle of a single point earthing system is the optimum solution. Two key requirements when setting up an effective earthing system are:

- Minimize the effects of impedance coupling between different circuits (i.e. when different signal currents flow through common impedance).
- Ensure that earth (ground) loops are not created (for example, do not tie the screen of a cable at two points to earth).

There are three types of earthing systems possible. These are illustrated in Figure 4.14 below. The series single point configuration is the most common; whilst the parallel single point configuration is the preferred approach with a separate earthing system for each circuit. In general, earthing can be classified into four different types of applications:

- Safety or power earth
- Low level signal (or telecommunications/instrumentation) earth
- High level signal (motor controls) earth
- Building earth

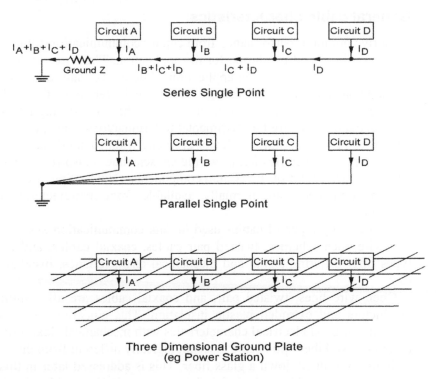

Figure 4.14
Various earthing configurations

4.5.5   Specific areas to focus on

- Always tie the shield of the cable down at one point to reduce the electrostatic field coupling.
- Twisted pair signal leads are preferable to non-twisted pair cables as they result in far less inductive noise.

- Ensure that all cables (even if both are data signal cables) cross at right angles.
- Multiple conductor cable should have an overall electrostatic shield.
- Keep high power cables as close together as possible to maximize the cancellation of disturbing magnetic fields.
- Terminate all unused wires and shields at one end of the cable.
- Terminate all electrostatic shields at one point only.
- Ensure isolation of electrostatic shields from ground at all points (once terminated at one end).
- At junction boxes, ensure that shield continuity is maintained even though the signal cables may be discontinued at this point.
- Ensure that signal cables avoid high noise areas.
- Do not assume that two signal cables of similar voltages will not interfere with each other electrically.
- It is important to tie all earths at one site to a common ground, so that if there is a severe earth fault (or lightning strike) all points rise to the same potential and no one is electrocuted.

4.6 Types of cables

4.6.1 General cable characteristics

To derive optimum performance from copper communication cables, the correct type and size of cable should be chosen for the application. The wire size must reflect the current carrying requirements for the application while the voltage rating should equal or exceed the anticipated circuit rated voltage. Mechanical strength of the cable must be addressed in order that the physical stresses imposed on the cable during installation and operation are acceptable, (and are not detrimental to terminations). Increased strength of cables can be afforded by using multiconductor grouping within a single jacket.

The type and specification of wiring chosen should also reflect the noise susceptibility and data rate criteria of the system components. (Note – Specific noise details relating to system components are normally available from manufacturers or suppliers of the equipment.)

Of the many types of cables used in data communication systems, common types are two wire open schemes, twisted pair cables, coaxial cables, and fiber optic cables. All types offer differing signal and mechanical characteristics, installation convenience and cost.

Open wire lines, twisted pair, and coaxial cables are all manufactured with copper conductors and extruded plastic insulation. This construction combines the important elements of good electrical characteristics with mechanical flexibility, ease of installation and low cost. Fiber optic cable is technologically different from the above and uses beams of light transmitted down a glass fiber. This is addressed later in this section. Aluminum conductors are seldom used for data communication cables because of their higher resistance and other physical limitations such as lack of flexibility.

For copper cables the resistance depends on the cross-sectional area of the conductor, measured in mm^2 and the length of the cable. The thicker the conductor, the lower the resistance, the lower the signal voltage drop, and the higher the current it can carry without excessive heating.

The signal voltage drop ($V_{drop} = I \times R$), depends on:

- The line current – dependent on the receiver input and transmitter output impedances
- The conductor resistance – dependent on wire size and length

For DC voltages and low frequency signals, the resistance of the conductor is the only major concern. The volt drop along the cable affects the magnitude of the signal voltage at the receiving end and in the presence of noise, affects the signal-to-noise ratio and the quality of the signal received.

As the frequency (or data transfer rates) increases, the other characteristics of the cable, such as capacitance and series inductance become important. The inductance and capacitance are factors that depend on how the cable is made, the quality of copper used, the number of shields and thickness of shields used, the number of fibers in the cable and the type of insulation materials used.

The resistance, inductance, and capacitance are distributed along the length of the cable and, at high frequencies, combine to present the effects of a low pass filter. The equivalent electrical circuit of a cable is illustrated in Figure 4.15 with these parameters shown distributed along the length of the cable.

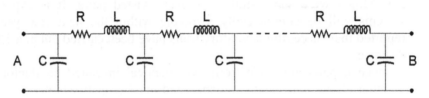

Figure 4.15
The main parameters of a data communications cable

To derive the correct performance for data communications, the correct type and size of cable should be chosen for the application. The following general rules apply to most data applications, although there are some new types of twisted pair cables that give good high frequency performance.

- Low data transfer rates → Low frequency type cables (e.g. twisted pair cables)
- High data transfer rates → High frequency type cables (e.g. coaxial cables, optic fiber or high quality twisted pair data cables)
- High noise environment cables → Shielded copper or optic fiber

Another important consideration when choosing a cable is its type of outer insulation and protection. For example, a cable may have the following options:
- A thin aluminum tape wound around the cable, under the plastic insulation, to provide a barrier against moisture ingress (moisture will permeate through plastic over a period of time) and provide an extra shield. This is often referred to as a moisture barrier.
- Steel armored outer for areas where protection is required against excessive heat, fire, or mechanical damage.
- Filled with a petroleum jelly between pairs to provide a good moisture barrier.
- A nylon coated outer to provide protection from rodents and provide a slippery jacket for ease of installation in conduits.
- Plenium cables, which are made with non toxic insulation, for installation in non vented areas (no poisonous fumes produced in fires).

4.6.2 Two wire open lines

Two wire open lines are the simplest forms of copper cable transmission media where each is insulated from the other and separated in free space. For digital transmission, the scheme is limited to transmission rates of up to 19 kbps over a range of 50 m and is susceptible to spurious noise. If an analog modulation scheme is used, such as frequency shift keying (FSK) then distances of several hundred kilometers are possible for pole mounted open wire systems. Speeds of up to 9600 baud are generally possible but they are still susceptible to noise.

4.6.3 Twisted pair cables

Twisted pair cables are the most economical solution for data transmission and allow for transmission rates of up to 1 Mbps on communication links of up to 300 m (longer distances with lower data transfer rates). Some new types of twisted pair cables (e.g. Twistlan) are suitable for up to 10 Mbps. Twisted pairs can be STP (shielded twisted pair) or UTP (unshielded twisted pair).

Tests have recently been carried out in a factory environment where speeds of 100 and 200 Mbit/s were successfully run over twisted pairs. It is expected that 500 Mbit/s systems will be commercially available within the next few years. It has also been reported that successful laboratory trials have been carried out at a 1 Gbps over very short distances.

Twisted pairs are made from two identical insulated conductors, which are twisted together along their length at a specified number of twists per meter, typically 40 twists per meter (12 twists per foot). The wires are twisted to reduce the effect of electromagnetic and electrostatic induction. An earth screen and/or shield is often placed around them to help reduce the electrostatic (capacitive) induced noise. An insulating PVC sheath is provided for overall mechanical protection. The cross-sectional area of the conductor will affect the IR loss, so for long distances thicker conductor sizes are recommended. The capacitance of a twisted pair is fairly low at about 10 to 50 pF/ft, allowing a reasonable bandwidth and an achievable slew rate.

For full-duplex digital systems using balanced differential transmission, two sets of screened twisted pairs are required in one cable, with individual and overall screens. A protective PVC sheath then covers the entire cable.

Due to the rapid increase during the 70s and 80s of twisted pair cables in data communications applications, the EIA developed a structured wiring system for unshielded twisted pair (UTP) cables. This structure provides a set of rules and standards for the selection and installation of UTP cables in data communications applications up to 100 Mbit/s.

The structure involves dividing UTP into five categories of application. These are listed below:

- CATEGORY 1 UTP   –   Low speed data and analog voice
- CATEGORY 2 UTP   –   ISDN data
- CATEGORY 3 UTP   –   High speed data and LAN
- CATEGORY 4 UTP   –   Extended distance LAN
- CATEGORY 5 UTP   –   Extended frequency LAN

The following table shows the electrical characteristics defined for the cables and connectors used with Categories 3, 4 and 5 UTP.

	Category 3 ≤ 10 MHz	Category 3 ≥ 10 MHz	Category 4 ≤ 10 MHz	Category 4 ≥ 10 MHz	Category 5 ≤ 10 MHz	Category 5 ≥ 10 MHz	Category 5 ≤100 MHz
Connector Attenuation (dB)	0.4	0.4	0.1	0.2	0.1	0.2	0.4
Cable Impedance (ohms)	100	100	100	100	100	100	100
Cable Attenuation (dB/km)	26 (1MHz) 98 (10MHz)	131	21 (1MHz) 72 (10MHz)	89 (16MHz) 102 (20MHz)	20 (1MHz) 6.6 (10MHz)	82 (16MHz) 92 (20MHz)	220
Patch Cord Impedance (ohms)	100	100	100	100	100	100	100
Patch Cord Attenuation (dB/km)	25 (1MHz) 98 (10MHz)	131	26 (1MHz) 98 (10MHz)	131	26 (1MHz) 98 (10MHz)	131	131

Table 4.4

The connection point of a landline into a building or equipment shelter is at the main distribution frame (MDF) or intermediate distribution frame (IDF).

In making data connections to modems, telemetry units or computer equipment, it is common to use withdrawable multiconductor connectors (e.g. 9-pin, 15-pin, 25-pin, 37- pin, 50-pin, etc). These connectors are usually classified as follows:

- The type, make or specification of the connector
- The number of associated pins or connections
- The gender (male or female)
- Mounting (socket or plug)

For example, the well known connector DB-25 SM specifies a D-type, 25 pin socket, male (with pins).

There are many different types of connectors used by computer manufacturers such as IBM, Hewlett Packard, Wang, Apple, etc and the various manufacturers of printers, radio equipment, modems, instrumentation, and actuators. The following is a selection of some of the more popular connectors:

- DB-9, DB-15, DB-25, DB-37, DB-50
- Amphenol 24-pin
- Centronics 36-pin
- Telco 50-pin
- Berg 50-pin
- RJ-11 4-wire
- RJ-11 6-wire
- RJ-45 8-wire
- DEC MMJ
- M/34 (CCITT V.35)
- M/50

There is also a wide range of DIN-type connectors (German/Swiss), IEC-type connectors (European/French), BS-type connectors (British) and many others for audio, video, and computer applications. With all of these, the main requirement is to ensure compatibility with the equipment being used. Suitable types of connectors are usually recommended in the manufacturer's specifications.

The DB-9, DB-25 and DB-37 connectors, used with the EIA standard interfaces such as RS-232, RS-422 and RS-485, have become very common in data communications applications. The interface standards for multidrop serial data communications, RS-422 and RS-485 do not specify any particular physical connector. Manufacturers who sell equipment, complying with these standards, can use any type of connector, but the DB-9, DB-25 (pin assignments to RS-530), DB-37 (pin assignments to RS-449 and sometimes screw terminals, have become common. Another connector commonly used for high speed data transmission is the CCITT V.35 34-pin connector.

4.6.4   Coaxial cables

Coaxial cables are discussed in detail in *Practical Radio Engineering and Telemetry for Industry*.

4.6.5   Fiber optics

The data transmission capability of fiber optic cables will satisfy any future requirement in data communications, allowing transmission rates in the gigabits per second (Gbps) range. There are many systems presently installed operating at approximately 2.5 Gbps. Commercial systems are becoming available that will operate up to 5 Gbps.

Fiber optic cables are designed for the transmission of digital signals and are therefore not suitable for analog signals. Although fiber optic cables are generally cheaper than coaxial cables (when comparing data capacity per unit cost), the transmission and receiving equipment together with more complicated methods of terminating and joining these cables, makes fiber-optic cable the most expensive medium for data communications. The cost of the cables themselves has halved since the late 1980s and is becoming insignificant in economic equation. It is worth noting that fiber optic technology has become more affordable over the last decade and this trend will continue in the future.

The main benefits of fiber optic cables are:
- Enormous bandwidth (greater information carrying capacity)
- Low signal attenuation (greater speed and distance characteristics)
- Inherent signal security
- Low error rates
- Noise immunity (impervious to EMI and RFI)
- Logistical considerations (light in weight, smaller in size)
- Total galvanic isolation between ends (no conductive path)
- Safe for use in hazardous areas
- No Crosstalk

4.6.6   Theory of operation

The principle of communications in fiber optic cables arises from the fact that light propagates through different media at different speeds (in the same manner as radio waves). When light moves from one media of a certain density to another of a different density, the light will change direction. This phenomena, is known as refraction.

It is possible to state the effectiveness of a medium to propagate light by expressing it as a ratio to an absolute reference; light traveling through a vacuum (3×10^8 m), i.e. speed of light in free space. This ratio is known as the refractive index and is calculated below.

$$n_1 = \frac{C_v}{C_1}$$

where:
C_v = Speed of light in vacuum (meters/sec)
C_1 = Speed of light in the other medium (meters/s)
n_1 = Refractive index for the other medium

In a typical fiber optic medium, light travels at approximately 2×10^8 meters/second. Therefore the refractive index is:

$$n_1 = \frac{3 \times 10^8}{2 \times 10^8} \text{ m/sec}$$
$$= 15$$

The optical medium is said to have a refractive index of 1.5.

Fiber optics can be explained by Willebrod Snell's law, which states that the ratio of the sine of the angle of incidence to the sine of the angle of refraction (Sin_i / Sin_r) is equal to the ratio of the speed of light in the two respective media (C_1/C_2). This is equal to a constant (K), which is a ratio of the refractive index of medium 2 to medium 1 (n_2/n_1). The formula appears below:

$$\frac{Sin_{\angle i}}{Sin_{\angle r}} = \frac{C_1}{C_2} = K = \frac{n_2}{n_1}$$

Fiber optic cables are manufactured with a core made of pure optical glass, surrounded by an optical cladding. The core and cladding are treated with an impurity so that their refractive indices are different. Figures 4.16 and 4.17 show the basic construction of the optic fiber. This construction principally allows the core to guide the light pulses to the receiver.

Because the refractive index of both the core and cladding are different, light entering the core at an acceptable angle of entry will propagate the length of the fiber without losing light through the cladding and beyond. The 'cone of acceptance' angles are the angles at which light moves through the cable. When light enters at an angle greater than this, it does not reflect from the cladding and is lost.

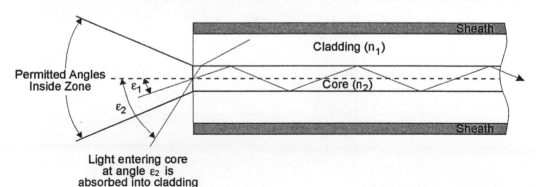

Figure 4.16
Optic fiber principles

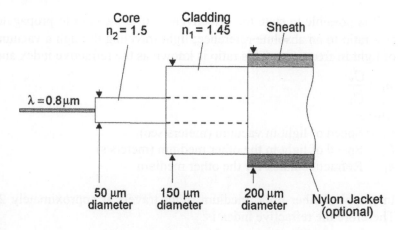

Figure 4.17
Typical optic fiber values (multimode fiber – step index)

The optic fiber therefore acts as a conduit (or wave-guide) for pulses of light generated by a light source. The light source is typically either an injection laser diode (ILD) or LED operating at wavelengths of 0.85, 1.2 or 1.5 μm (micrometers). The optic fiber is coated with a protective colored sheath to provide stability and allow easy identification.

4.6.7   Modes of propagation

Fiber types are generally identified by the number of paths that the light follows inside the fiber core called modes of propagation. There are two main modes of light propagation through an optic fiber, which give rise to two main constructions of fiber, multimode and monomode.

Multimode fibers are easier and cheaper to manufacture than monomode fibers. Their cores are typically 50 times greater than the wavelength of the light signal they will propagate. An LED transmitter light source is normally utilized with this type of fiber and can thus be coupled with less precision than would be necessary with an ILD.

With the wide aperture of the multimode fiber and its LED transmitter, it will send light in multiple paths (modes) toward the receiver as illustrated in Figure 4.18.

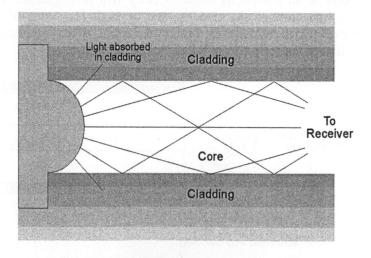

Figure 4.18
LED light source coupled to a multimode fiber (step index)

The light therefore takes many paths between the two ends as it reflects from the sides of the core/cladding of the fiber. As a result, the original sharp pulses at the sending end become distorted at the receiving end because the light paths arrive both out of phase and at different times causing a spreading of the original pulse shape.

The problem becomes worse as data rates increase. The effect is known as modal dispersion. The result of this effect is referred to as intersymbol interference.

Multimode fibers therefore have a limited maximum data rate (bandwidth) as the receiver can only differentiate between the pulsed signals at a reduced data rate. For slower data rates over short distances, multimode fibers are quite adequate. Speeds of up to 2–300 Mbit/s are readily available on multimode systems.

A further factor to consider with multimode fibers is the index of the fiber (how the impurities are applied in the core). The cable can be either graded index (more expensive but better performance) or step index (less expensive), refer to Figure 4.19. The type of index used affects the way in which the light waves reflect or retract off the walls of the fiber. Graded index cores focus on the modes as they arrive at the receiver and consequently improve the permissible data rate of the fiber.

The core diameters of multimode fibers typically range between 50 to 100 micrometers. The two most common core diameters are 50 and 62.5 μm.

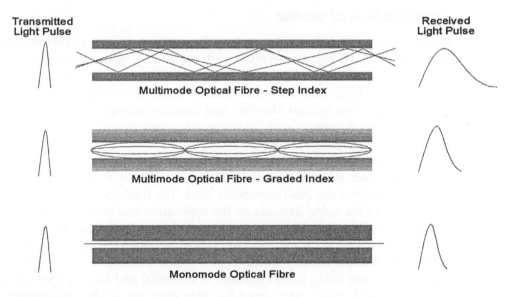

Figure 4.19
Optic fibers and their characteristics

Monomode or single mode fibers are expensive and difficult to manufacture. They allow only a single path or mode for the light to travel down the fiber with minimal reflections. ILDs are typically employed as light sources.

Monomode fibers do not suffer from major dispersion or overlap problems and permit a very high rate of data transfer over much longer distances. The fibers are much thinner than multimode fibers at approximately 5 to 10 micrometers.

Light from the source must be powerful and be aimed more precisely into the fiber to overcome any misalignment (hence the use of ILDs). The thin monomode fibers have implications when splicing and terminating as they are more difficult to work with and expensive to install.

A typical application for monomode installations would be high capacity links used by telephone companies where the traffic volume makes it necessary to install a communications medium with a large bandwidth.

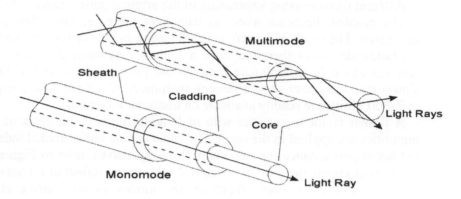

Figure 4.20
Monomode and multimode optic fibers

4.6.8   Specification of cables

Optic fibers are specified based on their diameters. A fiber specified as 50/150 means one with a core of 50 micrometers and with a cladding diameter of 150 micrometers. The most popular sizes of multimode fibers are 50/125, mainly in Europe and 62.5/125, mainly in Australia and the USA. Another layer, called the outer coating, provides an external protection against abrasion and shock. Coatings can range from 200 to 1000 micrometers in diameter. Very often, the cable specification includes the diameter of the coating; for example 50/150/250.

To provide additional mechanical protection, the fiber is often placed inside a loose, but stiffer, outer jacket which adds thickness and weight to the cable. Cables made up with several fibers are most commonly used. The final sheath and protective coatings on the outside of the cable depends on the application and where the cable will be used. A strengthening member is normally placed down the center of the cable to give it longitudinal strength. This allows the cable to be pulled through a conduit or hung from a power pole without causing damage to the fibers. The tensile member is normally steel or Kevlar, the latter being more common. In industrial and mining applications, fiber cores are often placed inside cables used for other purposes, such as trailing power cables for large mining, stacking, or reclaiming equipment.

Experience has shown that optic fibers will break two or three times in a 25-year period. In general, the incremental cost of extra fiber cores in cables is not very high when compared to overall costs (including installation and termination costs). Therefore, it is often worthwhile specifying extra cores as spares or for future use.

4.6.9   Joining cables

In the early days of optic fibers, the connections and terminations were a major problem. Largely, this has improved but connections still require a great deal of care to avoid signal losses that will affect the overall performance of the communications system.
There are three main methods of connecting optic fibers:
- Mechanical – where the fibers are fitted with plug/socket arrangements and simply plug together.

- Chemical – where the two fibers are fitted into a barrel arrangement with an epoxy glue in it. They are then heated in an oven to set the glue.
- Fusion splicing – where the two fibers are welded together with heat.

To overcome the difficulties of termination, fiber optic cables can be provided by a supplier in standard lengths such as 10 m, 100 m or 1000 m with the ends cut and finished with a mechanical termination ferrule that allows the end of the cable to slip into a closely matching female socket. This enables the optic fiber to be connected and disconnected as required. The mechanical design of the connector forces the fiber into a very accurate alignment with the socket and results in a relatively low loss. Similar connectors can be used for in-line splicing using a double-sided female connector.

Although the loss through this type of connector can be an order of magnitude greater than the loss of a fused splice, it is much quicker and requires no special tools or training. Unfortunately, mechanical damage or an unplanned break in a fiber requires special tools and training to repair and re-splice. One way around this problem is to keep spare standard lengths of pre-terminated fibers that can quickly and easily be plugged into the damaged section. The techniques for terminating fiber optics are constantly being improved to simplify these activities.

4.6.10 Limitations of cables

On the negative side, the limitations of fiber optic cables are as follows:
- The cost of source and receiving equipment is relatively high.
- It is difficult to switch or tee-off a fiber optic cable, so fiber optic systems are most suitable for point-to-point communications links.
- Techniques for joining and terminating fibers (mechanical and chemical) are difficult and require precise physical alignment. Special equipment and specialized training are required.
- Equipment for testing fiber optic cables is different and more expensive from the traditional methods used for electronic signals.
- Fiber optic systems are used almost exclusively for binary digital signals and are not really suitable for analog signals.

4.7 Privately owned cables

At a majority of factories and industrial sites there will probably already exist a large network of communications cables. The cables may be high quality copper data cables, standard voice grade twisted pair copper cables, or fiber optic cables. Some computer local area networks may be using coaxial cables.

Integrating a telemetry system into these cabling systems can be very cost effective if there is enough capacity on the cable network for the telemetry links and if RTUs are located near existing cable termination points.

4.7.1 Telephone quality cables

If standard telephone quality twisted pair cables are to be used, then generally, the connection of modems to these cables or when the equipment is available, the use of data transceivers operating on the RS-422 or RS-485 standards will be required. The speeds at which these can operate will depend upon the transmission distance, the quality of the cable (its age) and the noise on the cable. Generally, this will be limited to approximately 100 kbps for distances up to 1 km.

If there is a shortage of cable pairs, then new multicore twisted pair telephone cables can be run and any excess pairs could be used for connecting extra telephones, remote computer terminals, telemetry equipment for expansion at a later date or other unrelated applications. This makes running new telephone cables a reasonably attractive option.

The connection of modems to landlines and the RS-422 and RS-485 standards are discussed in Chapter 6.

4.7.2 Data quality twisted pair cables

Although it would be preferable to use data quality twisted pair cables around a factory or industrial site for transmission of telemetry data, the expense involved usually makes it impractical. Generally, only just enough data quality cables are run for computer terminals and there are rarely any excess pairs. However, if there are, it is recommended that good use be made of them. They are less susceptible to noise and will carry higher data rates over longer distances than voice quality lines. Data speeds of 200 kbps over 1 km or 500 kbps over 300 m are possible using RS-422 or RS-485 standard transceivers.

4.7.3 Local area networks (LANs)

At a site, it may be possible to connect some telemetry RTUs into an existing LAN. The network may be running on fiber optic cable such as the fiber distributed data interface (FDDI) or 10 Base F Ethernet standards, or on coaxial cable such as the 10BASE5 Ethernet standard or on twisted pair such as the token ring standard.

To interface into the network, the RTU will require an appropriate LAN interface card with an RS-232 connection for interface to the telemetry equipment. Connection may be made through a media access unit (MAU) of a LAN.

Appropriate software would have to be provided at the RTUs and master site to provide an appropriate protocol for accessing and sending data on and off the network.

There are many different types of LANs in use and thousands of different products for providing access to the networks. If this approach is to be undertaken then advice should be sought from the supplier of the LAN as to which are the most appropriate interface products.

To determine if using the LAN is an appropriate solution for implementing a telemetry system, the following should be given careful consideration:

- The cost of using LAN products compared to using radio or copper cable systems.
- The effect that the telemetry data traffic will have on the existing LAN traffic. It may load the system down severely and cause major delays and annoyance to users (this will depend on the size of the telemetry system to be installed and the available data capacity on the LAN).
- Minimum access times required by the telemetry system, compared to maximum wait time for access to the LAN. Some LANs have long or indeterminate wait times during heavy traffic periods.
- All RTUs should be very close to the LAN cabling.
- The cost involved in extending the LAN cabling (including bridges) if RTUs are at a significant distance from the existing cabling.

4.7.4 Multiplexers (bandwidth managers)

It may be appropriate to use a data channel on an existing statistical multiplexer, that is being used with a bandwidth management system of a data communication network around a factory or industrial site, for accessing certain RTUs. These are relatively simple to implement. Circuit cards are purchased for the multiplexer (if they are not already installed) that has interfaces compatible with the RTUs and master site. This is normally RS-232, for speeds up to 19.2 kbps. Higher speeds up to 64 kbps are readily available. There are no response time problems with this type of installation but there are normally only a small number of multiplexers of this sort installed on one site.

Figure 4.21 illustrates an example installation on a site where use is made of a LAN, statistical multiplexers and data and voice cabling for accessing all RTUs.

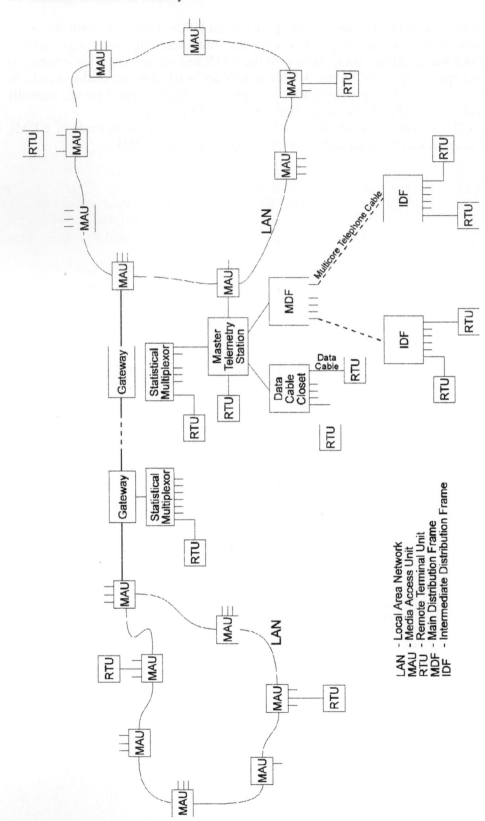

LAN - Local Area Network
MAU - Media Access Unit
RTU - Remote Terminal Unit
MDF - Main Distribution Frame
IDF - Intermediate Distribution Frame

Figure 4.21
Telemetry installation in a factory or industrial site using LAN, cable, and multiplexer access

4.7.5   Assessment of existing copper cables

The first step in setting up a telemetry system over existing copper cables is to determine the present condition of these cables. One method is to place a test pair of telemetry units over the lines that are to be used and run them for 24 hours to determine how they perform. This is generally not recommended because although this pair of telemetry units may be operating satisfactorily, they may be operating very close to the failure point where data begins to corrupt and when another two units are put in with a slightly reduced adjusted tolerance (or the weather conditions change) then the system may fail. In addition, it does not give an accurate indication of the condition of the cable link itself.

A good methodology for assessing the condition of a cable requires two measurements. Firstly, a noise and distortion meter should be placed across the terminated line and the noise level measured. Then data should be sent from a telemetry unit at the far end of the cable and the level measured again. With these measurements the signal to noise ratio should be calculated.

The second part of the cable test should consist of placing a bit error rate tester on the line in a looped back mode, and measuring the number of errors that occur in a minimum period of 30 minutes. The longer the test is performed the more the result will reflect the actual state of the cable.

These tests should be performed during the normal operating conditions of the factory or industrial site so that all relevant noise and interference will be present.

Normally when copper cables start to exhibit high levels of noise because of age, there is little that can be done to fix them. An exception may be when there is a problem with an old joint or IDF. In this case, the joint can normally be cut out and replaced. In most other cases, new cables will be required.

4.8   Public network provided services

When the RTUs of a telemetry system are to lie outside the immediate boundaries of an industrial site, it is often necessary to lease the communications services of a public telephone network provider, to get reliable communications access to them. Quite often, this may be the only option as the sites may be too distant for line of sight radio or microwave or may be obstructed by mountains or buildings or there may be problems obtaining suitable frequencies in built up urban and industrial environments where the frequency spectrum is often very crowded.

The main providers of infrastructure and services are generally referred to as carriers. There are a number of smaller suppliers who buy large blocks of capacity from the main carriers and then resell smaller pieces of this capacity. These are mostly data services and are quite often only available in central business districts.

IDC recommend that once you have decided on the type of service you require, it is worth contacting the various carriers and suppliers to determine if they can provide this type of service to your master and RTU locations. Then a cost comparison should be carried out between the different service providers.

The telephone networks of the carriers are often referred to as the public switched telephone network or just the PSTN. The data networks are referred to as the public switched data network or PSDN.

The services provided by the smaller carriers and suppliers use fundamentally the same technology as those provided by the larger carriers. This chapter will examine a typical range of services from a larger carrier. The knowledge will then be directly applicable to other company's services. Quite often, the only difference between the services from different companies is the marketing name used for the service and the level of personal service and maintenance provided by the company.

The following sections will firstly examine analog services available and then the various digital services available.

4.9 Switched telephone lines

4.9.1 General

The most common PSTN connection that will be used for telemetry is the standard telephone connection that is fitted into every house.

For this type of connection the telemetry system would work on an exception-reporting basis where the RTU would automatically dial up the master when an alarm occurred or there was a change of status at the RTU. The other method of operation over a standard telephone line would be where the master dialing the RTU or the RTU dialing the master at a predetermined regular interval.

Using a standard switched telephone line limits the type of telemetry implementation to non-critical applications. As continuous contact is not kept with the RTU, when the RTU fails it is not detected until the next time that the master site dials it. There is also the dialing delay from when the alarm occurs until when the master answers the call. If the lines to the master are busy, it could take a significant number of redials to get through. The advantages of using a switched line are:

- There is easy access to a telephone line in most urban locations and some rural locations (depending generally on population density).
- Easy and quick installation.
- A good range of cheap modems is available for connection to the lines.
- The costs involved are the initial installation costs and from then on only the annual rental costs and call costs. Compared to other services these costs are relatively small.

It should be noted that this is a switched service. That is, it sends the call through a switching machine (a branch exchange) that is designed to only hold the line up for a finite period of time (the statistical average length of a telephone call is three minutes). The call may go through a series of relays or electronic switches.

In countries where the costing of calls is not on a timed basis but on a per call basis, users have tried to dial up an RTU and then leave the line open as a permanent communications link in an effort to avoid the rental costs of a more expensive dedicated link. Although this is certainly possible, it is not recommended. The main problem is that telephone companies are required to regularly reload software and do maintenance on the exchanges and in doing so, disconnect all the lines from the exchange. This will therefore drop out any through connected modem calls.

4.9.2 Technical details

The telephone company's equipment is referred to as the exchange end of a switched line while the telephone/modem end is referred to as the subscriber.

The standard switched telephone line is a two wire 600-ohm circuit. It allows full duplex communications (i.e. signals can flow in both directions simultaneously) over the two wires. At the exchange the two wire circuit will connect to a hybrid circuit and be converted from two wire to four wire. It will then transverse through the telephone network in either digital or analog form, depending on whether the exchange is digital or analog.

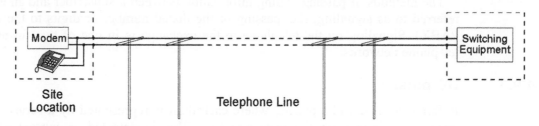

Figure 4.22
Illustration of a telephone connection

CCITT have recommended international performance standards for bandwidth, attenuation distortion, phase shift, SNR and levels to and from the exchange and subscriber. Most countries conform to the limits of the standards. Figure 4.23 illustrates a common configuration of levels. This illustration shows an allowance for a line loss between the exchange and the subscriber of approximately 8 dB. Therefore the subscriber would be transmitting at approximately –1 dBm and receiving at approximately –8 dBm. Attenuation pads are normally inserted into the line at the exchange to get the levels out of the hybrid to approximately +4 dB and –13 dBm in all circuits. There is approximately 0.5 dB loss through the switch and 3.5 dB loss through the hybrid.

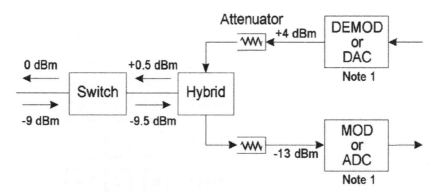

Note 1 - Demodulation and modulation required for frequency division multiplexing of **analog exchanges.**

- Digital to **Analog Converters** (DAC) and Analog to Digital Converters (ADC) are required for digital exchanges.

Figure 4.23
Audio levels at exchange

The exchange feeds a low current 50 volt supply down each exchange line to power the telephone. The standard telephone connection is referred to as a ring/loop connection. This is because the line will loop in the direction of the telephone to the exchange and ring in the direction of the exchange to the telephone (or modem). When the telephone user wishes to make a call the handset is lifted which puts a loop across the line indicating to the exchange to expect some dialing digits. The modem does the same when it wishes to make a call. The exchange will send ring pulses to the subscriber end to make the telephone ring. When the telephone handset is lifted and the line is looped, the exchange stops sending ringing pulses and connects the voice circuit through to the telephone.

The methods of passing dialing information between a subscriber and an exchange are referred to as signaling, (i.e. passing of the dialed number or digits to the exchange or PABX). Signaling by the telephone to the exchange is in one of two forms for analog telephone networks:

4.9.3   DC pulses

Referred to as decadic pulsing where each digit is represented by a series of on/off (or open/close) breaks on the exchange line. This is detected as current pulses at the exchange. The breaks are expected at the exchange in a sequence with approximately one tenth of a second between each break in a digit. For example, two consecutive breaks represent the digit 2, nine consecutive breaks the digit 9, ten consecutive breaks the digit 0 and so on.

4.9.4   Dual tone multifrequency – DTMF

This system consists of a three by four matrix of audio frequency tones, where the telephone or modem to the exchange sends two tones for each digit dialed. Each column or row represents a different frequency. This method allows significantly faster dialing and therefore more efficient use of the telephone system. The standard keypad is illustrated in Figure 4.24.

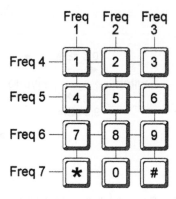

Figure 4.24
Standard DTMF keypad

DTMF dialing is used with the modern digital or part digital exchanges and decadic dialing is used with the older relay set type exchanges (electromechanical).

4.10   Analog tie lines

4.10.1   Introduction

The main limitation with using switched line services is that the user does not have continuous real time access to the RTU and is required to dial the RTU's telephone number to access it. Therefore, they can only be used for non-critical applications. For critical applications, we require permanent links.

In telecommunications terminology a permanent hard-wired fixed link between two locations through a PSTN is referred to as a tie line.

This type of link is dedicated to one user and is not switched through any switching equipment in the exchange (with exception to ISDN tie lines discussed in Section 4.14) and therefore no other subscriber can get access to it. The line is available to each user 24 hours a day 12 months of the year (assuming there are no line failures).

Tie lines are also sometimes referred to as:

- Leased lines, or
- Permanent lines, or
- Dedicated lines, or
- Private circuits.

Any of these terms refer to the same type of connection.

There are both analog and digital tie lines. The first type to be examined will be analog tie lines. They are available from the carrier in a number of different forms. Each of these will be discussed in the following sections.

4.10.2 Four wire E&M tie lines

The four wire E&M tie line is normally the top of the range analog tie line available. These are sometimes referred to as 'conditioned lines' because equalizers, filters and amplifiers are used to improve the quality of the lines. The subscriber is provided with a six wire circuit at both ends of the link. The six wires are allocated to the following functions:

- Two wires for transmit signals
- Two wires for receive signals
- One wire for E signaling
- One wire for M signaling

The E&M signaling terminology is derived from the description ear and mouth signaling (or in some circles 'exchange' and 'multiplex'). These wires simply transmit and receive decadic pulses for dialing purposes. Instead of using loop/ring signally and decadic pulses for dialing over the same two wires carrying audio signals as is used in switched circuits, the dialing information is sent over the E&M lines. The M line is for signaling out (to the remote end) and the E line is for signaling in (from the remote end).

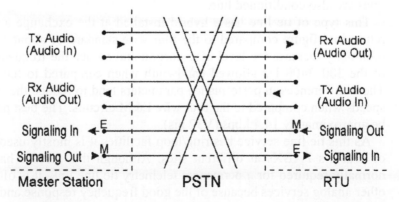

Figure 4.25
Wiring configuration for four wire E&M circuit

Signaling is achieved by placing an earth on the M lead and receiving an earth from the E lead.

Most E&M leads will accept pulse rates of around 10 pulses per second. Therefore, fast automatic dialing can be achieved and the E&M leads may be used for transmitting separate slow speed additional signaling data.

There are a number of benefits with having separate transmit and receive pairs. Firstly there are no hybrids in the circuit and therefore the return loss is greatly improved. Return loss for a switched two wire line will be approximately 12 dB compared to approximately 45 dB for the four wire circuit.

Secondly, there is less distortion and crosstalk over the rated operating frequency range (300–3400 Hz) and finally there is a better frequency response over the specified frequency range. These features together usually allow higher data rates using standard analog modem modulation techniques.

The four wire E&M tie may be slightly more sophisticated than is needed for the average telemetry system. As the tie line will be permanently connected from master to RTU the E&M signaling will most likely not be required. The reason such a circuit may be used is for high speed modem data (up to 38.4 kbps – V42 bis), or where a low noise high integrity analog link is used.

This type of circuit is normally used for connecting the user's private automatic telephone exchanges (PABX) at two separate locations (a PABX is the telephone system installed at a customer's premises).

The premium quality lines are quoted as meeting the CCITT performance standard M.1020. IDC suggest that typical performance figures would be:
- Premium: 600 ohms impedance
- Nominal transmit level of –10 dBm
- Insertion loss of 0 to 10 dB
- SNR of around 30 dB
- ± 2 dB to ± 4 dB level variation across a 300 to 3000 Hz bandwidth

4.10.3 Two wire signaling tie line

The next level of tie line service down is the dedicated two wire line between two sites and which has the same facilities as the standard switched line. This service is fundamentally a two wire tie line with ring/loop type signaling on each end. Therefore the two wires carry all the transmit, receive and signaling information. Two wire signaling tie lines are also conditioned lines.

This type of tie line has a hybrid installed at the exchange and therefore has a lower return loss figure compared to the four wire standard of about 12 dB. This also tends to minimally increase the level of distortion and reduce the frequency response at the limits of the 300–3400 Hz allowed bandwidth when compared to the four wire E&M circuit. These differences in performance parameters tend to reduce the maximum attainable data speeds when compared to the four wire E&M circuits. The best possible data speed would be approximately 14.4 kbps (V32 bis).

As this tie line service has ring/loop facilities it is mostly used for network connection of a user's PABXs at different sites. Although this service has more features than is normally required for a permanent telemetry tie line link, it is often used in preference to other analog services because of the good frequency response and low insertion loss.

These lines are also specified to CCITT performance standard M1020 with similar expected performance figures as quoted for four wire E&M lines (except for return loss.)

4.10.4 Four wire direct tie lines

Four wire direct tie lines are four wire point-to-point connections. They have no signaling facilities and are generally poor brothers to the four wire E&M signaling tie lines. They can be used for non-critical point-to-point connections of relatively slow speed modems. These are sometimes referred to as 'unconditioned lines' as minimal equalization is used on the lines. There are generally different grades of lines available, from low quality to high quality.

This type of tie line is often used for telemetry and alarm applications where data rates up to 4800 bit/s (for standard quality) and 9600 bit/s (from premium quality) are more than adequate.

As can be seen from the performance parameters, this type of service is not of a guaranteed high quality when compared to the voice link services. The standard quality lines are quoted as meeting the CCITT performance standard M.1040. IDC suggest that typical performance figures would be:

- Standard: 600 ohms impedance
- Nominal transmit level of −10 dBm
- Insertion loss of 10 to 20 dB
- SNR of around 20 dB
- ± 6 dB to ± 16 dB level variation across a 300 to 3000 Hz bandwidth

4.10.5 Two wire direct tie lines

Two wire direct tie lines are simple two wire point-to-point connections. As with the four wire direct tie line it has no signaling facilities and is generally of a poorer quality than the two wire signaling tie line. They can be used for non-critical point-to-point telemetry applications where relatively slow speed data are required (as with four wire tie lines).

These lines are also specified to CCITT performance standard CCITT M.1040 with similar performance figures as quoted for four wire direct lines.

4.11 Analog data services

Service providers will often provide a point-to-point or point-to-multipoint data service over analog tie lines. The service will sometimes include the network terminating units (in this case modems) and sometimes just a data quality analog tie line.

Most countries around the world refer to these as DATEL services. The following is a generic description of the types of services provided. You will have to check with your local service provider to determine if the services as described here are available to you.

With many suppliers at each exchange all these lines are taken into a remote line access and test unit. This unit allows the exchange to subscriber lines to be tested for noise and other inherent problems, and to carry out end-to-end loop backs for bit error rate testing. Each access and text exchange unit is connected by permanent lines to a central test site, where data testing staff and fault reporting facilities are situated. Here a central site operator can gain access to any circuit countrywide.

The services are generally classified according to three parameters:

- The type of line used for connection to the exchange, i.e. switched or dedicated.
- The speed of operation.
- As point-to-point or point-to-multipoint operation.

A feature of this service is the facility to provide a point-to-multipoint network for polled master-to-multiple slave operation.

4.11.1 Introduction

DATEL is an international term for data transmission services over standard analog bandwidth lines. The service can be provided over switched or leased lines and includes the provision of a modem at each end. End-to-end maintenance is also included as part of the service.

The DATEL service is provided on analog network lines. The service is classified according to three parameters:

- The type of line used, i.e. switched or dedicated
- The speed of operation
- As point-to-point or point-to-multipoint operation.

4.11.2 Point-to-point configuration

The DATEL service can be configured for point-to-point operation where there is a single line between a master and an RTU. The line connection can be with either switched or dedicated lines. A typical DATEL point-to-point configuration is shown in figure 4.26.

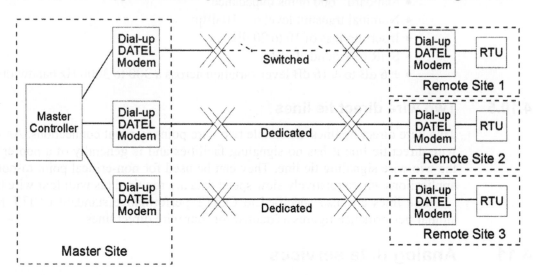

Figure 4.26
DATEL point-to-point configuration

4.11.3 Point-to-multipoint

It is also possible to have a point-to-multipoint configuration in which a number of connections are established between a master unit and a number of RTUs on leased lines (not on switched lines). The connections are established on four wire leased lines and can provide significant cost savings over having to connect a number of point-to-point links. The disadvantages of this configuration are, firstly, that if the link to the master fails then communications with all RTUs is lost. Secondly, data throughput for this system will be significantly slower than for a separate point-to-point DATEL service to each RTU.

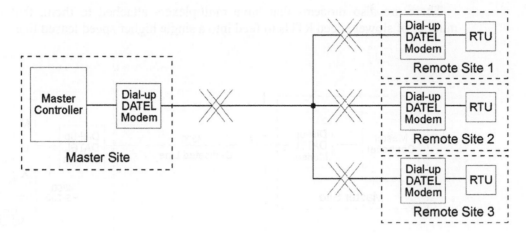

Figure 4.27
DATEL point-to-multipoint configuration

This feature of DATEL to provide a point-to-multipoint network is useful for polled master-to-multiple slave operation. The multiple branching points are provided by the central test centers at each exchange. This type of network infrastructure is suitable for point-to-multipoint telemetry systems where the RTUs are located at remote locations. The cost of setting up a multipoint network through the DATEL facility is significantly cheaper than implementing individual leased lines from the central site to each of the remote locations.

4.11.4   Digital multipoint

If at one of the locations there are a number of RTUs within close proximity, it is possible to wire them into a single multipoint connection using RS-232 connection. A device referred to as a multiple junction unit (MJU) is used, which allows a number of RS-232 connections into a single modem on a multipoint basis.

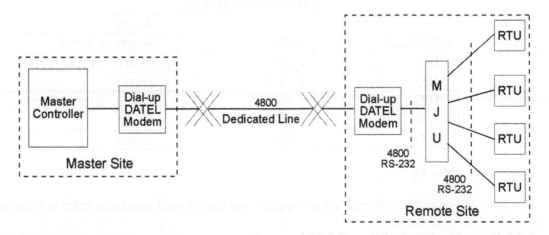

Figure 4.28
DATEL point-to-multipoint digital connection for same location

There are also modems that have multiplexers attached to them, that will allow a number of slower speed RTUs to feed into a single higher speed leased line.

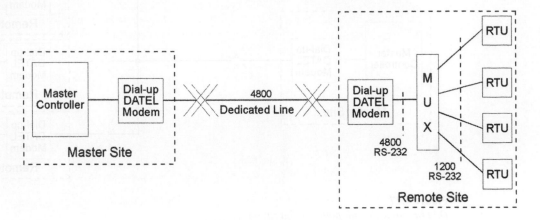

Figure 4.29
DATEL point-to-multipoint connection with multiplexer

4.11.5   Switched network DATEL service

Switched network DATEL services operate through the public switched telephone network (PSTN). They are sometimes also referred to as data exchange line (DXL) services. Depending on the service type, there will be a requirement for one or two exchange line connections. Each exchange line will require a telephone to be connected in parallel with the modem.

Note that all DATEL modems have RS-232 (CCITT V24 & V28) standard digital interfaces. The following table lists the range of DATEL switched network services generally available.

Speed Bits	Relevant CCITT Recommendation	Mode of Operation
300	V.21	Async
600/1200	V.23	Async / Sync
1200	V.22	Start / Stop / Sync
2400	V.26 bis	Sync
2400	V.22 bis	Start / Stop / Sync
4800	V.27 ter	Sync
9600	V.29	Sync

Table 4.5

For further information on modems and the relevant standards, refer to Chapter 5.

4.11.6   Dedicated line DATEL service

These services operate on exclusive end-to-end tie line connections, in point-to-point or point-to-multipoint configuration. They are also referred to as DATEL dedicated line (DDL) services. No telephone is required to be connected across the dedicated line.

The following information lists the range of DATEL services available.

- Low speed service: This is for services in the 300 to 1200 bit/s range, over any distance.
- Local area service: For data services within the same exchange area.
- Short distance services: For data services between locations that are off different exchanges, where the exchanges are less than 10 km apart.
- Long distance services: Data services between locations, where exchanges are greater than 10 km apart.
- High speed services: Data services over wide band analog circuits.

4.11.7 Additional information

In some countries, at each exchange all DATEL lines are taken into a remote line access and test unit, referred to as RATS (remote access and test systems). This unit allows the exchange to subscriber lines to be tested for noise and other inherent problems, and to carry out end-to-end loop backs for BER testing. Each RATS is connected by permanent lines to a central test site, where data testing staff and fault reporting facilities are situated. Here a central site operator can gain access to any circuit countrywide. Charges for leased line services have the following components:

- Rental or purchase of the modems for both ends of the link (either from the carrier or a private supplier)
- Monthly rental of the leased line from the exchange to the subscriber (fixed price at all locations)
- Monthly rental of the leased lines between exchanges if required (charged on a per kilometer basis)

4.12 Digital data services

4.12.1 General

The services discussed so far have all been the transmission of data over analog mediums using modems to access the lines. The next step is to use a digital transmission medium from the remote units to the exchange.

Many service providers worldwide have established dedicated digital data networks specifically designed to carry digital tie line services, for speeds from 1200 baud to 64 kbps.

4.12.2 Service details

A dedicated digital transmission service for point-to-point and point-to-multipoint links is referred to as the digital data service (DDS). The DDS offers data rates of 2400, 4800, 9600 and 48 kbps for synchronous, leased line services with similar functionality to the DATEL services.

A dedicated digital transmission service for point-to-point links can offer data speeds of 2400, 4800, 9600, 48 k, 64 k and multiples of 64 k up to 2 Mbps. The data connections are synchronous on leased digital lines and some leased analog lines for the slower speeds.

Multiplex units are placed at the exchanges where all the different speed links from the users are multiplexed together for transmission between exchanges. The multiplexing is done onto E1 PCM 30 channel links in Europe, Asia, Africa, or T1 PCM 30 channel link in the USA.

The DDS network is a dedicated digital data network that has been specially designed to carry digital transmission. It is not part of the PSTN. It is a sophisticated data network where all links are monitored from a central network administration center from where faults can be quickly detected BER testing can be carried out and services restored quickly when outages occur.

DDS services are generally more expensive than normal leased analog lines or DATEL services. This is because it is a digital service, which provides a higher guaranteed availability than equivalent analog services and because of the advanced network monitoring and administration facilities. For this reason it is rarely used in telemetry applications. The exception being where a very high quality data link of high reliability is required for an application where the data link is a critical part of the process.

DDS is generally the 'top of the line' data communications service offered for point-to-point links. Services such as ISDN and X.25 do not have the same testing and administration facilities as DDS. DDS generally offer three types of interfaces:

- X.21 bis for synchronous speeds up to 9600 bps and asynchronous speeds up to 2400 bps
- V.35 (RS449) for data speeds of 48 k bps on wide band analog data lines
- X.21 for synchronous services from 2400 to 64 k bps.

The DDS services are charged on the basis of speed of the circuit and on the distance between end users. There is a fixed charge for terminals hanging off the one exchange and the cost for users hanging off different exchanges depends on the distance between the exchanges.

All services provided are X.21 bis interface standard, operating in synchronous mode. Also available through this service are multiplexer facilities for a site to bring a number of slower speed data links into a high-speed data link. There are also facilities to bring 2 Mbps into a site through the DDN.

Note that data into the DDN network must be phase and rate synchronized with the DDN output data clock. This clock is provided at the DDN network terminating units (NTU) by the carrier. If an RTU is at a significant distance from the NTU, there may be a delay in the data from the RTU, which will make it out of phase with DDN data. Therefore, special protocols are required to realign the incoming data from the RTU with the DDN clock.

There is also a service that is offered to users in the metropolitan area of major cities, referred to as the digital metropolitan service (DMS). Fundamentally, it is a basic DDS service, offered at the speeds of 2400, 4800 and 9600 bps in a point-to-point or point-to-multipoint configuration.

4.13 Packet switched services

4.13.1 Introduction

An important method now commonly used for transporting data through public switched data networks is with the use of data packets.

Data is stored at the terminal until a specified number of bytes have been collected and then the data is formed into packets and sent off into a packet switched network. A system will be continually sending these packets of data into the network as it produces data for transmission.

In the packet switched network the packets of data are stored at the site they entered until there is free transmission capacity available to forward the data to the next packet network site. Because of this store and forward action, there is no guarantee that the packets of data will arrive at the receiver terminal in the same order they were sent. Also there will be a significant delay (in the realms of data communications) from when the packet is sent to when it is received.

The exact time of this delay is difficult to predict. It will depend mostly on the amount of traffic that is flowing through the network at the time and the link capacity between packet network sites. For this reason packet switched data services are generally not used for time critical applications.

Suppliers of these services publish delay objectives for their system and show the normal mean delay achieved, but do not provide guaranteed delay times. The mean delay for an average system including call setup and clear down times is about 1.5–2.0 seconds. The worst case delay is about 3 seconds.

The largest packet of data that can be sent is approximately 4000 bytes (i.e. 4000 characters). But on average, packets will be around 140 bytes in length.

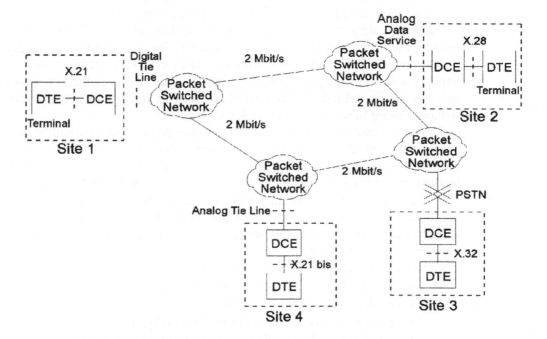

Figure 4.30
Packet switched network connections

Figure 4.30 illustrates a possible implementation of a packet switched network connection of four sites. X.25 is the CCITT standard that describes the protocol of how packets of data are framed, formatted, and then sent to the packet switched network (often referred to as the public switched packet data network (PSPDN).

Normally a dedicated digital data link or a dedicated analog data link is used to provide access from the sites to the packet network.

At site one, the link between the X.25 terminal and the network access unit (DCE) is a synchronous digital link operating to the CCITT X.21 standard. The link to the network is via a dedicated digital tie line link. The DCE is often referred to as a packet assembler/disassembler (PAD).

At site two, an asynchronous digital link between the terminal (DTE) and the network access unit (DCE) required the use of the CCITT X.28 connection. In this case an analog dedicated line is provided back to the terminal.

At site three, access to the packet switched network is made through the normal analog PSTN network. Therefore, a special interface standard is required for the terminal (DTE) and the modem (DCE) to correctly access the packet switched network. In this case, the CCITT X.32 interface standard is used.

Site 4 is a distant site and the terminal (DTE) has been for many years used with analog synchronous modems. The physical interface standard CCITT X.21 bis is used to allow X.25 terminals to directly interface to an analog network access unit (DCE).

Details of these services now follow.

4.13.2   X.25 service

X.25 requires a unit to be installed at all customer sites referred to as a packet assembler dissembler (PAD). This receives the data from the DTE, assembles it into appropriate data packet sizes, and then sends them into the PSDN. Generally, telecommunication service providers do not provide PADs as the type of PAD that is used depends on the features required and the type of DTE it is to be connected to.

Generally, data transfer is achieved by setting up the call between the two ends, sending the data and then closing the call down. Because there is no physical connection made between the two ends this is referred to as a 'virtual circuit'. It is possible with the Saponet service to setup a 'permanent virtual circuit' between the two ends where once a call has been established it is possible to keep it open and transmit data at any time without having to re-establish the link each time.

The data transit time through the network depends on link availability and on the traffic on the network at the time of transmission. The aim according to the standard is to provide a maximum exchange to exchange transit time of 1.25 seconds at 90% traffic loading on the network. The DTE to DTE transit time will vary from 1 to 3 seconds.

The X.25 service caters for all standard synchronous speeds from 2400 to 64000 bps inclusive (DTE to PAD speed). There are many options available with the X.25 service that should be examined if this service were to be used. Information is available from your local service provider.

The costs associated with X.25 are as follows:

- The rental cost of the analog or digital leased line
- The data speed of the connection
- The rental or purchase cost of the NTU (PAD)
- Call charges – broken down into the following elements:
    - Packet count charge
    - Call duration charge
    - Call attempt charge

(The first being the most significant component)

The rates applied to these vary with the time of day and week at which they are sent, and on the message priority applied to the packet.

4.13.3   X.28 services

The X.25 services were designed for use by larger organizations, and not for smaller occasional users with low data transfer requirements. HENCE, X.28 was designed to service this end of the market. X.28 have the following features:

- Operates in asynchronous mode only
- Data speeds of 1200 and 2400 bps

- Can be used with dedicated analog or digital lines
- Can be used with dial up lines
- Data transfer is generally in ASCII
- Packet assembly and disassembly is carried out by a multiport PAD located at the exchange and not at the subscribers premises
- Operation is half duplex only (X.25 can be full duplex)
- There is no automatic error detection and recovery in the protocol as there is in X.25

The charges associated with X.28 are the same as for X.25 except there are no priority messages available.

4.13.4　X.32 services

X.32 was developed for the requirement where the user requires the full functionality of X.25 but only requires occasional access to the network. It can also be used as a backup to an X.25 service if the leased line fails.

X.32 defines the interface between the packet switched network and a packet mode DTE that gains access via the PSTN. The communications protocol is the same as the one used by X.25 (HDLC). Transmission between the DTE and the network is in full duplex synchronous mode using any standard V-series dial up modem. The two speeds presently supported are 2400 and 9600 bps using V.22 bis and V.32 modems.

The charges associated with X.32 are the same as X.25 except there are no leased line rentals, just switched line rentals.

X.32 is a good option for remote telemetry only requiring occasional access and where the system ties back into a larger packet switched system.

4.13.5　Frame relay

Frame relay is a relatively new concept in packet switching, providing very high data rates. It is a very simplified X.25 type service, which does away with sequence numbering, frame acknowledgment, and automatic retransmission and concentrates on getting data through fast to the right place and in the right order.

It uses a shortened version of the HDLC protocol. The concept is based on using one or more ISDN standard channels. Therefore, the slowest speed of operation is 64 kbps. Errors that occur in the network are not accounted for and errored packets are simply discarded. Error handling is left up to the end point terminals.

The charging for frame relay is very similar to X.25. Each circuit requires a leased 64 kbps digital line or ISDN line access.

4.14　ISDN

ISDN is an acronym for integrated services digital network. It was a recommendation that emerged from the Consultative Committee for International Telephone and Telegraphs (CCITT), an international standards organization in the late 1970s. The concept was to provide a truly open digital communications standard that would provide digital communication to all telecommunication users from households to large multinational organizations.

ISDN is provided to the customer at two levels. The first is referred to as basic rate access (BRA). BRA consists of 192 kb linked into the customer's premises. This link is broken down into 2 × 64 kbps channels, referred to as the B channels, and a 16 kbps channel referred to as the D channel. The B channel can be used for voice or data and the D channel is used for signaling for voice circuits and/or slow speed data.

The second level of service is referred to the primary rate service (PRA). PRA consists of 30 × 64 kbps channels (i.e. 30 × B channels) for voice and data, 1 × 64 kbps signaling channel for signaling and data and 1 × 64 kbps clocking channel.

These services are functionally the same as the analog switched telephone services into the PSTN but the major difference being that they are purely digital switched lines. Dedicated 64 kbps tie lines can be established through the ISDN switched digital network if required (referred to as semi-permanent lines). Now, because of the high speeds involved with ISDN, they are of little relevance to telemetry applications. But over the next 10 or so years as digital services begin to replace analog services and the prices of switched digital technology drop, they will become more relevant to telemetry applications.

In the simplest context, ISDN is a digital form of the existing plain old telephone system (POTS). Germany is considered the leading country in the introduction of ISDN and there are now digital exchanges in all capital cities and some country towns. California, UK and France would closely follow in levels of implementation. A large group of exchange numbers is available and because ISDN overlays the existing analog network there are numerous junctions between the two systems so that telephone calls originating in ISDN can migrate to the POTS network in a manner that is quite transparent to the user.

The public telephone carrier will install a terminal adapter near to where the service is required and the user can connect his approved ISDN device. This may be a computer data line, a digital PABX or telephone, a video conferencing system or a group 4 facsimile machine that can transmit a copy machine quality facsimile to a similar machine in about five seconds. All of these devices are able to dial up a compatible machine via ISDN and calls can be as required or setup as a semi-permanent circuit with favorable tariff arrangements for long-term use. Figure 4.31 shows a general arrangement of a basic ISDN network.

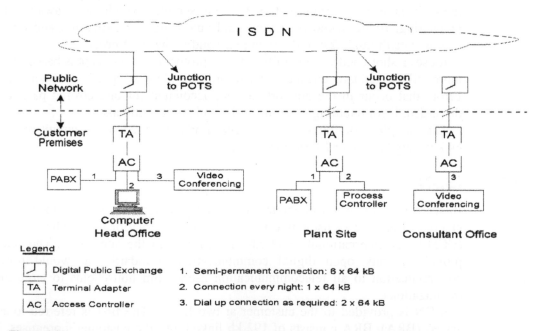

Figure 4.31
Basic ISDN network

One interesting application of ISDN allows it to be used as a backup to a normal data link running over a private network or via a public carrier which may operate as a data link between a plant process controller and a central computer. The equipment is connected to the normal circuit via an adapter unit at each end. A second connection from the adapter goes to an ISDN terminal adapter at each end. The ISDN is normally unused, however, if the adapters detect a loss of data over the normal link, they will quickly and automatically establish an ISDN connection for the duration of the failure. Figure 4.32 shows how such a system would operate.

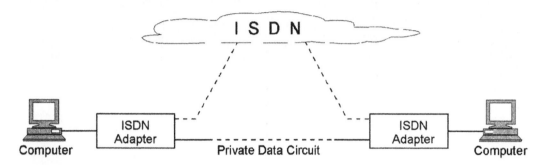

Figure 4.32
ISDN backup to a data link

4.15 ATM

Asynchronous transfer mode (ATM) has been hailed as the standard of data and telecommunications transmission that will eventually replace all other standards (but in the fast changing telecommunications industry, one should be very careful using statements like that). It will allow the simultaneous transmission of voice, data, and video over the one network.

Standards are still under development but the majority of them have been ratified, with everything due for completion by mid 1996. Virtually all data communication products' vendors have ATM equipment available

ATM is an extension of ISDN referred to as broadband ISDN (B-ISDN). At this stage, minimum planned connection speeds are 45 Mbps.

ATM data links use a high-speed form of packet switching called cell switching which carries data in 53 byte cells with 5 byte cell headers. Very fast cell switching is used to create virtual circuits. This provides users with the advantages of circuit switching and network bandwidth to suit the particular application.

ATM technology will provide networks, which have the characteristics of today's LANs, but with each user having the required bandwidth for each particular application. Similarly ATM WANs can be implemented. ATM technology is likely to replace existing LAN and WAN technology in the long term.

5

Local area network systems

5.1   Introduction

Local area networks (LANs) are about sharing information and resources. To enable all the nodes on the network to share this information, they must be connected by some transmission medium. The method of connection is known as the network topology.

The nodes need to share this transmission medium in such a way as to allow nodes access to the medium and minimize disruption of an established sender. The main methods of this media access control will be discussed and their effects on system performance investigated.

A LAN is a communications path between one or more computers, file-servers, terminals, workstations, and various other intelligent peripheral equipment, which are generally referred to as devices or hosts. A LAN allows access to devices to be shared by several users, with full connectivity between all stations on the network. A LAN is usually owned and administered by a private owner and is located within a localized group of buildings.

The connection of a device into a LAN is made through a node. A node is any point where a device is connected and each node is allocated a unique address number. Every message sent on the LAN must be prefixed with the unique address of the destination. All devices connected to a node also watch for any messages sent to its own address on the network. LANs operate at relatively high speed (Mbps range and upwards) with a shared transmission medium over a fairly small geographical (i.e. local) area.

In a LAN, the software controlling the transfer of messages among the devices on the network must deal with the problems of sharing the common resources of the network without conflict or corruption of data. Since many users can access the network at the same time, some rules must be established on which devices can access the network, when and under what conditions. These rules are covered under the general subject of media access control. The rules that apply depend on the structure of the network (i.e. the rules are different for a star, ring or bus topology).

When a node has access to the channel to transmit data, it sends the data within a packet, which includes, in its header, the addresses of both the source and the destination. This allows each node to receive or ignore data on the network. A frame is often used to indicate the packet sent (or the message transmitted). This is derived from the idea that the transmitter frames the data with a preamble and postamble consisting of special packaging characters.

5.2   Network topologies

The way the nodes are connected to form a network is known as its topology. There are many topologies available but they form two basic types: logical or physical.

A logical topology defines how the elements in the network communicate with each other, and how information is transmitted through a network. The different types of media-access methods, discussed in Section 5.3, determine how a node gets to transmit information along the network. In a broadcast topology, all the information broadcast goes to every node within the amount of time it actually takes a signal to cover the entire length of cable. This time interval limits the maximum speed and size for the network. In a ring topology, each node hears from exactly one node and talks to exactly one other node. Information is passed sequentially, in an order determined by a predefined process. A polling or token mechanism is used to determine who has transmission rights, and a node can transmit only when it has this right.

A physical topology defines the wiring layout for a network. This specifies how the elements in the network are connected to each other electrically. This arrangement will determine what happens if a node on the network fails. Physical topologies fall into three main categories – bus, star, and ring topology. Combinations of these can be used to form hybrid topologies to overcome weaknesses or restrictions in one or other of these three component topologies.

5.2.1   Bus topology

A bus refers to both a physical and a logical topology. As a physical topology, a bus describes a network in which each node is connected to a common single communication channel or 'bus'. This bus is sometimes called a backbone, as it provides the spine for the network. Every node can hear each message packet as it goes past. Logically, a passive bus is distinguished by the fact that packets are broadcast and every node gets the message at the same time. Transmitted packets travel in both directions along the bus and need not go through the individual nodes, as in a point-to-point system. Rather, each node checks the destination address that is included in the message packet to determine whether that packet is intended for the specific node. When the signal reaches the end of the bus, an electrical terminator absorbs the packet energy to keep it from reflecting back again along the bus cable, possibly interfering with other messages already on the bus. Each end of a bus cable must be terminated, so that signals are removed from the bus when they reach the end. In a bus topology, nodes should be far enough apart so that they do not interfere with each other. However, if the backbone bus cable is too long, it may be necessary to boost the signal strength using some form of amplification, or repeater. The maximum length of the bus is limited by the size of the time interval that constitutes 'simultaneous' packet reception. Figure 5.1 illustrates the bus topology.

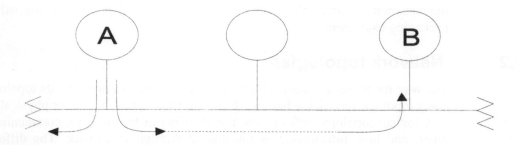

Figure 5.1
Bus topology

5.2.2   Bus topology advantages

Bus topologies offer the following advantages:
- A bus uses relatively little cable compared to other topologies and arguably has the simplest wiring arrangement.
- Since nodes are connected by high impedance tappings across a backbone cable, it's easy to add or remove nodes from a bus. This makes it easy to extend a bus topology.
- Architectures based on this topology are simple and flexible.
- The broadcasting of messages is advantageous for one-to-many data transmissions.

5.2.3   Bus topology disadvantages

These include the following:
- There can be a security problem, since every node may see every message – even those that are not destined for it.
- Diagnosis / troubleshooting (fault-isolation), can be difficult, since the fault can be anywhere along the bus.
- There is no automatic acknowledgment of messages, since messages get absorbed at the end of the bus and do not return to the sender.
- The bus cable can be a bottleneck when network traffic gets heavy. This is because nodes can spend much of their time trying to access the network.

5.2.4   Star topology

A star topology is a physical topology in which multiple nodes are connected to a central component, generally known as a hub. The hub of a star generally is just a wiring center; that is, a common termination points for the nodes, with a single connection continuing from the hub. In some cases, the hub may actually be a file server (a central computer that contains a centralized file and control system), with all its nodes attached directly to the server. As a wiring center, a hub may, in turn, be connected to the file server or to another hub. All signals, instructions, and data going to and from each node must pass through the hub to which the node is connected. The telephone system is doubtless the best known example of a star topology, with lines to individuals coming from a central location. There are not many LAN implementations that use a logical star topology. The low impedance ARCnet networks are probably the best examples. However, you will see that the physical layout of many other LANs look like a star topology even though they are considered to be something else. Examples of star topologies are shown in Figure 5.2.

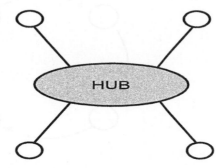

Figure 5.2
Star topology

5.2.4.1 Star topology advantages

- Troubleshooting and fault isolation are easy.
- It is easy to add or remove nodes and to modify the cable layout.
- Failure of a single node does not isolate any other node
- The inclusion of a central hub allows easier monitoring of traffic for management purposes.

5.2.4.2 Star topology disadvantages

- If the hub fails, the entire network fails. Sometimes a backup central machine is included, to make it possible to deal with such a failure.
- A star topology requires a lot of cable.

5.2.5 Ring topology

A ring topology is both a logical and a physical topology. As a logical topology, a ring is distinguished by the fact that message packets are transmitted sequentially from node to node, in a predefined order, and as such it is an example of a point-to-point system. Nodes are arranged in a closed loop, so that the initiating node is the last one to receive a packet. As a physical topology, a ring describes a network in which each node is connected to exactly two other nodes. Information traverses a one-way path, so that a node receives packets from exactly one node and transmits them to exactly one other node. A message packet travels around the ring until it returns to the node that originally sent it. In a ring topology, each node can acts as a repeater, boosting the signal before sending it on. Each node checks whether the message packet's destination node matches its address. When the packet reaches its destination, the destination node accepts the message, then sends it back to the sender, to acknowledge receipt. As you will see later in this chapter, since ring topologies use token passing to control access to the network, the token is returned to sender with the acknowledgment. The sender then releases the token to the next node on the network. If this node has nothing to say, the node passes the token on to the next node, and so on. When the token reaches a node with a packet to send, that node sends its packet. Physical ring networks are rare, because this topology has considerable disadvantages compared to a more practical star-wired ring hybrid, which is described later.

Figure 5.3 shows some examples of the ring topology.

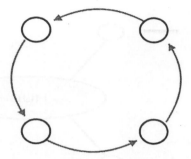

Figure 5.3
Ring topology

5.2.5.1 Ring topology advantages

- A physical ring topology has minimal cable requirements
- No wiring center or closet needed
- The message can be automatically acknowledged
- Each node can regenerate the signal

5.2.5.2 Ring topology disadvantages

- If any node goes down, the entire ring goes down.
- Diagnosis/troubleshooting (fault isolation) is difficult because communication is only one-way.
- Adding or removing nodes disrupts the network.
- There will be a limit on the distance between nodes.

5.3 Media access methods

A common and important method of differentiating between different LAN types is to consider their media access methods. Since there must be some method of determining which node can send a message, this is a critical area that determines the efficiency of the LAN. There are a number of methods, which can be considered, of which the two most common in current LANs are the contention method and the token passing method. You will become familiar with these as part of your study of LANs, although some of the other methods will also be briefly discussed.

5.3.1 Contention systems

The basis is first-come-first-served media access method. This operates in a similar manner to polite human communication. We listen before we speak, deferring to anyone who already is speaking. If two of us start to speak at the same time, we recognize that fact and both stop, before starting our messages again a little later. In a contention-based access method, the first node to seek access when the network is idle will be able to transmit. Contention is at the heart of the carrier sense; multiple access; collision detection (CSMA/CD) access method used in the IEEE 802.3 and the original Ethernet networks.

Let us now discuss this access method in more detail. The carrier sense component involves a node wishing to transmit a message listening to the transmission media to ensure there is no 'carrier' present. The length of the channel and the finite propagation delay means that there is still a distinct probability that more than one transmitter will attempt to transmit at the same time, as they both will have heard 'no carrier'. The collision detection logic ensures that more than one message on the channel simultaneously will be detected and transmission, from both ends, eventually stopped. The system is a probabilistic system, since access to the channel cannot be ascertained in advance.

5.3.2 Token passing

Token passing is a deterministic media-access method in which a token is passed from node to node, according to a predefined sequence. A token is a special packet, or frame, consisting of a signal sequence that cannot be mistaken for a message. At any given time, the token can be available or in use. When an available token reaches a node, that node can access the network for a maximum predetermined time, before passing the token on.

This deterministic access method guarantees that every node will get access to the network within a given length of time, usually in the order of a few milliseconds. This is in contrast to a probabilistic access method (such as CSMA/CD), in which nodes check for network activity when they want to access the network, and the first node to claim the idle network gets access to it. Because each node gets its turn within a fixed period, deterministic access methods are more efficient on networks that have heavy traffic. With such networks, nodes using probabilistic access methods spend much of their time competing to gain access and relatively little time actually transmitting data over the network. Network architectures that support the token-passing access method include token bus, ARCnet, FDDI, and token ring.

5.3.2.1 Token-passing process

To transmit, the node first marks the token as 'in use', and then transmits a data packet, with the token attached. In a ring topology network, the packet is passed from node to node, until the packet reaches its destination. The recipient acknowledges the packet by sending the message back to the sender, who then sends the token on to the next node in the network.

In a bus topology network, the next recipient of a token is not necessarily the node that is nearest to the current token-passing node. Rather, the next node is determined by some predefined rule. The actual message is broadcast on to the bus for all nodes to 'hear'. For example, in an ARCnet or token bus network, the token is passed from a node to the node with the next lower network address. Networks that use token passing generally have some provision for setting the priority with which a node gets the token. Higher level protocols can specify that a message is important and should receive higher priority.

5.4 IEEE 802.3 Ethernet

Ethernet uses the CSMA/CD access method discussed in Section 5.4.1. This gives a system, which can operate with little delay, if lightly loaded, but the access mechanism can fail completely if too heavily loaded. Ethernet is widely used commercially because the network hardware is relatively cheap and produced in vast quantities. Because of its probabilistic access mechanism, there is no guarantee of message transfer and messages cannot be prioritized. This means that critical messages (such as alarms) on the one hand and unimportant messages on the other hand have the same priority in accessing the network. It is becoming widely used industrially despite these disadvantages. However, what has made the Ethernet LAN approach so popular is simply cost. It is often less than 10% of the price of the competing token-passing philosophy, which is generally preferred by the SCADA manufacturers, as being more reliable, guaranteeing consistent performance regardless of traffic conditions.

5.4.1 Ethernet types

The IEEE 802.3 standard defines a range of cable types that can be used for a network based on this standard. They include coaxial cable, twisted pair cable and fiber optic cable. In addition, there are different signaling standards and transmission speeds that can be utilized. These include both baseband and broadband signaling and speeds of 1 Mbps and 10 Mbps. The standard is continuing to evolve, and this manual will look at fast Ethernet (100 Mbps) and gigabit Ethernet (1000 Mbps) later in this chapter.

The IEEE 802.3 standard documents (ISO 8802.3) support various cable media and transmission rates up to 10 Mbps as follows:

- 10BASE-2 – thin wire coaxial cable (0.25 inch diameter), 10 Mbps, single cable bus
- 10BASE-5 – thick wire coaxial cable (0.5 inch diameter), 10 Mbps, single cable bus
- 10BASE-T – unscreened twisted pair cable (0.4 to 0.6 mm conductor diameter), 10 Mbps, twin cable bus
- 10BASE-F – optical fiber cables, 10 Mbps, twin fiber bus
- 1BASE-5 – unscreened twisted pair cables, 1 Mbps, twin cable bus (obsolete)
- 10BROAD-36 – cable television (CATV) type cable, 10 Mbps, broadband (obsolete)

5.4.2 10Base5 systems

This is a coaxial cable system and uses the original cable for Ethernet systems, generically called 'Thicknet'. It is a coaxial cable, of 50 ohm characteristic impedance, and yellow or orange in color. The naming convention for 10Base5: means 10 Mbps; baseband signaling on a cable that will support 500 meter segment lengths. It is difficult to work with and so cannot normally be taken to the node directly. Instead, it is laid in a cabling tray etc and the transceiver electronics (medium attachment unit, MAU) is installed directly on the cable. From there, an intermediate cable, known as an attachment unit interface (AUI) cable, is used to connect to the NIC. This cable, which can be a maximum of 50 meters long, compensates for the lack of flexibility of placement of the segment cable. The AUI cable consists of 5 individually shielded pairs – two each (control and data) for both transmit and receive; plus one for power.

The MAU connection to the cable can be made by cutting the cable and inserting an N-connector and a coaxial Tee or more commonly by using a 'bee sting' or 'vampire' tap. This mechanical connection clamps directly over the cable. Electrical connection is made via a probe that connects to the center conductor and sharp teeth, which physically puncture the cable sheath to connect to the braid. These hardware components are shown in Figure 5.4.

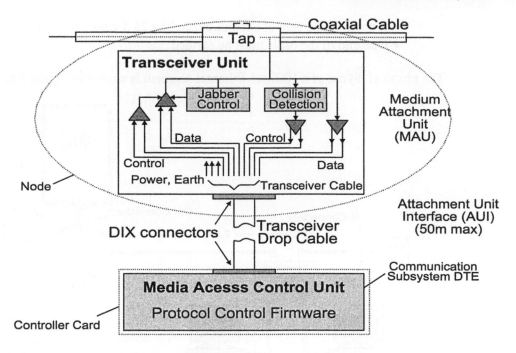

Figure 5.4
10Base5 hardware components

The location of the connection is important to avoid multiple electrical reflections on the cable and the Thicknet cable is marked every 2.5 meters with a black or brown ring to indicate where a tap should be placed. Fan out boxes can be used if there are a number of nodes for connection, allowing a single tap to feed each node as though it was individually connected. The connection at either end of the AUI cable is made through a 25 pin D-connector, with a slide latch, often called a DIX connector after the original consortium.

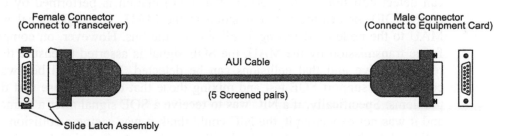

Figure 5.5
AUI cable connectors

There are certain requirements if this cable architecture is used in a network. These include:

- Segments must be less than 500 meters in length to avoid signal attenuation problems
- No more than 100 taps on each segment i.e. not every connection point can support a tap
- Taps must be placed at integer multiples of 2.5 meters
- The cable must be terminated with a 50 ohm terminator at each end

- It must not be bent at a radius exceeding 25.4 cm or 10 inches
- One end of the cable must be earthed

The physical layout of a 10Base5 Ethernet segment is shown in Figure 5.6.

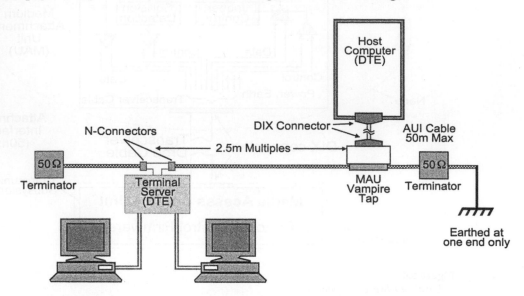

Figure 5.6
10Base5 Ethernet segment

The Thicknet cable was extensively used as a backbone cable until recently (1995), but 10BaseT and fiber are becoming more popular. Note that when a MAU (tap) and AUI cable is used, the on board transceiver on the NIC is not used. Rather, there is a transceiver in the MAU and this is fed with power from the NIC via the AUI cable. Since the transceiver is remote from the NIC, the node needs to be aware that the termination can detect collisions if they occur. This confirmation is performed by a signal quality error (SQE), or heartbeat, test function in the MAU. The SQE signal is sent from the MAU to the node on detecting a collision on the bus. However, on completion of every frame transmission by the MAU, the SQE signal is asserted to ensure that the circuitry remains active and that collisions can be detected. You should be aware that not all components support SQE test and mixing those that do with those that don't can cause problems. Specifically, if a NIC was to receive a SQE signal after a frame had been sent and it was not expecting it, the NIC could think it was seeing a collision. In turn, as you will see later in the manual, the NIC will then transmit a jam signal.

5.4.3   10Base2 systems

The other type of coaxial cable Ethernet networks is 10Base2 and often referred to as 'Thinnet' or sometimes 'thinwire Ethernet'. It uses type RG-58 A/U or C/U with a 50 ohm characteristic impedance and of 5 mm diameter. The cable is normally connected to the NICs in the nodes by means of a BNC T-piece connector, and represents a daisy chain approach to cabling. Connectivity requirements include:

- It must be terminated at each end with a 50 ohm terminator
- The maximum length of a cable segment is 185 meters and NOT 200 meters
- No more than 30 transceivers can be connected to any one segment
- There must be a minimum spacing of 0.5 meters between nodes

- It may not be used as a link segment between two 'Thicknet' segments
- The minimum bend radius is 5 cm

The physical layout of a 10Base2 Ethernet segment is shown in Figure 5.7.

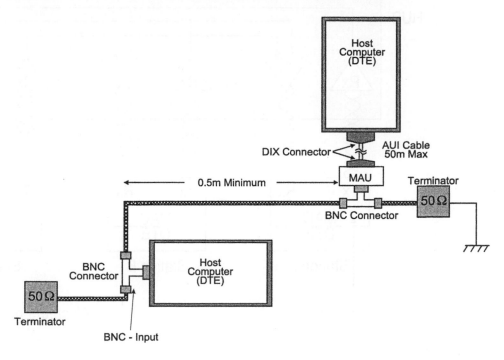

Figure 5.7
10Base2 Ethernet segment

The use of Thinnet cable was, and remains, very popular as a cheap and relatively easy way to set up a network. However, there are disadvantages with this approach. A cable fault can bring the whole system down very quickly. To avoid such a problem, the cable is often taken to wall connectors with a make–break connector incorporated. The connection to the node can then be made by 'fly leads' of the same cable type. It is important to take the length of these fly leads into consideration in any calculation on cable length. There is also provision for remote MAUs in this system, with AUI cables making the node connection, in a similar manner to the Thicknet connection.

5.4.4   10BaseT

The 10BaseT standard for Ethernet networks uses AWG24 unshielded twisted pair (UTP) cable for connection to the node. The physical topology of the standard is a star, with nodes connected to a wiring hub, or concentrator. Concentrators can then be connected to a backbone cable that may be coaxial or fiber optic. The node cable can be category 3 or category 4 cable, although you would be well advised to consider category 5 for all new installations. This will allow an upgrade path as higher speed networks become more common and given the small proportion of cable cost to total cabling cost, will be a worthwhile investment. The node cable has a maximum length of 100 meters; consists of two pairs for receive and transmit and is connected via RJ-45 plugs. The wiring hub can be considered as a local bus internally, and so the topology is still considered as a logical bus topology. Figure 5.8 shows schematically how the 10BaseT nodes are interconnected by the hub.

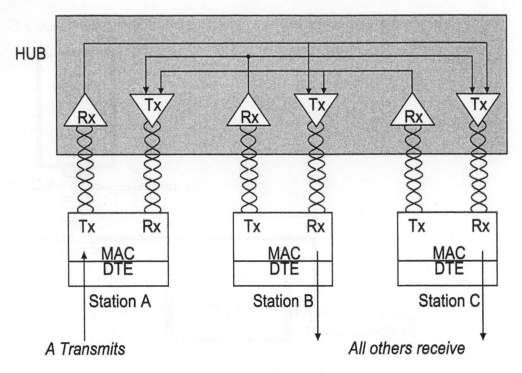

Figure 5.8
Schematic 10BaseT system

Collisions are detected by the NIC and so an input signal must be retransmitted by the hub on all output pairs. The electronics in the hub must ensure that the stronger retransmitted signal does not interfere with the weaker input signal. The effect is known as far end crosstalk (FEXT) and is handled by special adaptive crosstalk echo cancellation circuits.

The standard has become increasingly popular for new networks, although there are some disadvantages that should be recognized:

- The cable is not very resistant to electrostatic electrical noise, and may not be suitable for all industrial environments
- While the cable is inexpensive, there is the additional cost of the associated wiring hubs to be considered
- The node to hub cable distance is limited to 100 m

Advantages of the system include:

- Intelligent hubs are available that can determine which spurs from the hub receive information. This improves on the security of the network – a feature that has often been lacking in a broadcast, common media network such as Ethernet
- Flood wiring can be installed in a new building, providing many more wiring points than are initially needed, but giving great flexibility for future expansion. When this is done, patch panels – or punch down blocks – are often installed for even greater flexibility.

5.4.5   10BaseF

This standard, like the 10BaseT standard, is based on a star topology using wiring hubs. The actual standard has been delayed by development work in other areas and was ratified in September 1993. It consists of three architectures. These are:

- 10BaseFL

  The fiber link segment standard that is basically a 2 km upgrade to the existing fiber optic inter repeater link (FOIRL) standard. The original FOIRL as specified in the 802.3 standard was limited to a 1 km fiber link between two repeaters, with a maximum length of 2.5 km if there are 5 segments in the link. Note that this is a link between two repeaters in a network and cannot have any nodes connected to it

- 10BaseFP

  A star topology network based on the use of a passive fiber optic star coupler. Up to 33 ports are available per star and each segment has a maximum length of 500 m. The passive hub is completely immune to external noise and is an excellent choice for noisy industrial environments.

- 10BaseFB

  A fiber backbone link segment in which data is transmitted synchronously. It is designed only for connecting repeaters, and for repeaters to use this standard, they must include a built in transceiver. This reduces the time taken to transfer a frame across the repeater hub. The maximum link length is 2 km, although up to 15 repeaters can be cascaded, giving great flexibility in network design.

5.4.6   10Broad36

This architecture, while included in the 802.3 standard, is no longer installed as a new system. This is a broadband version of Ethernet, and uses a 75 ohm coaxial cable for transmission. Each transceiver transmits on one frequency and receives on a separate one. The Tx/Rx streams require a 14 MHz bandwidth and an additional 4 MHz is required for collision detection and reporting. The total bandwidth requirement is thus 36 MHz. The cable is limited to 1800 meters because each signal must traverse the cable twice, so the worst case distance is 3600 m. It is this figure that gives the system its nomenclature.

5.4.7   1Base5

This architecture, while included in the 802.3 standard, is no longer installed as a new system. It is hub based and uses UTP as a transmission medium over a 500 meter maximum length. However, signaling is 1 Mbps, and this means special provision must be made if it is to be incorporated in a 10 Mbps network. It has been superseded by 10BaseT.

5.4.8   Collisions

You should recognize that collisions are a normal part of a CSMA/CD network. The monitoring and detection of collisions is the method by which a node ensures unique access to the shared medium. It is only a problem when there are excessive collisions. This reduces the available bandwidth of the cable and slows the system down while retransmission attempts occur. With real-time industrial control networks the time delay incurred in such retransmissions is of real concern. Collisions can be reduced by keeping the segment length short and ensuring the traffic loading is light. A utilization of less than 5% is desirable on real-time systems to minimize collisions.

5.5 MAC frame format

The basic frame format for an 802.3 network is shown below. There are eight fields in each frame, and they will be described in detail.

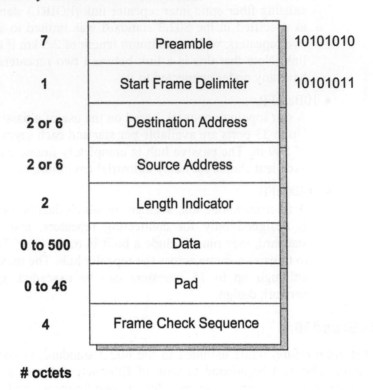

# octets		
7	Preamble	10101010
1	Start Frame Delimiter	10101011
2 or 6	Destination Address	
2 or 6	Source Address	
2	Length Indicator	
0 to 500	Data	
0 to 46	Pad	
4	Frame Check Sequence	

Figure 5.9
MAC frame format

Preamble

This field consists of 7 octets of the data pattern 10101010. It is used by the receiver to synchronize its clock to the transmitter.

Start frame delimiter

This single octet field consists of the data 10101011. It enables the receiver to recognize the commencement of the address fields.

Source and destination address

These are the physical addresses of both the source and destination nodes. The fields can be 2 or 6 octets long, although the six octet standard is the most common. The six octet field is split into two three octet blocks. The first three octets describe the block number to which all NICs of this type belong. This number is the license number and all cards made by this company have the same number. The second block refers to the device identifier, and each card will have a unique address under the terms of the license to manufacture. This means there are 2^{48} unique addresses for Ethernet cards.

There are three addressing modes that are available. These are:

- Broadcast – the destination address is set to all 1s or FFFFFFFFFFFF.
- Multicast – the first bit of the destination address is set to a 1. It provides group restricted communications.
- Individual, or point-to-point – first bit of the address set to 0, and the rest set according to the target destination node.

Length

A two octet field that contains the length of the data field. This is necessary since there is no end delimiter in the frame.

Information

The information that has been handed down from the LLC sublayer.

Pad

Since there is a minimum length of frame of 64 octets (512 bits or 576 bits if the preamble is included) that must be transmitted to ensure that the collision mechanism works, the pad field will pad out any frame that does not meet this minimum specification. This pad, if incorporated, is normally random data. The CRC is calculated over the data in the pad field. Once the CRC checks OK, the receiving node discards the pad data, which it recognizes by the value in the length field.

FCS

A 32-bit CRC value that is computed in hardware at the transmitter and appended to the frame. This is checked by the receiving node to verify the frame integrity.

5.6 High-speed Ethernet systems

Although Ethernet with over 200 million installed nodes world-wide is the most popular method of linking computers on a network, its 10 Mbps speed is too slow for very data intensive or real-time applications.

From a philosophical point of view there are several ways to increase speed on a network. The easiest, conceptually, is to increase the bandwidth and allow faster changes of the data signal. This requires a high bandwidth medium and generates a considerable amount of high frequency electrical noise on copper cables, which is difficult to suppress. The second approach is to move away from the serial transmission of data on one circuit to a parallel method of transmitting over multiple circuits at each instant. A third approach is to use data compression techniques to enable more than one bit to be transferred for each electrical transition. A fourth approach used with 1000 Mbps gigabit Ethernet is to operate circuits full duplex, enabling simultaneous transmission in both directions.

All of the three approaches are used to achieve 100 Mbps fast Ethernet and 1000 Mbps gigabit Ethernet transmission on both fiber optic and copper cables using the current high speed LAN technologies.

5.6.1 Cabling limitations

Typically, most LAN systems use coaxial cable, shielded (STP), unshielded twisted pair (UTP), or fiber optic cables. The capacitance of the coaxial cable imposes a serious limit to the distance over which the higher frequencies can be handled. Consequently, high-speed LAN systems do not use coaxial cables.

The unshielded twisted pair is obviously popular because of ease of installation and low cost. This is the basis of the 10BaseT Ethernet standard. The category 3 cable enables us to achieve only 10 Mbps while category 5 cables can attain 100 Mbps data rates, while the four pairs in the standard cable enable several parallel data streams to be handled.

As we have seen fiber optic cables have enormous bandwidths and excellent noise immunity so are the obvious choice for high-speed LAN systems.

5.7   100 Base-T (100Base-TX, T4, FX, T2)

5.7.1   Fast Ethernet overview

This is the preferred approach to 100 Mbps transmission, which uses the existing Ethernet MAC layer with various enhanced physical media dependent (PMD) layers to improve the speed. These are described in the IEEE 802.3u and 802.3y standards as follows:

IEEE 802.3u defines three different versions based on the physical media:

- 100Base-TX, which uses two pairs of category 5, UTP, or STP
- 100Base-T4, which uses four pairs of wires of category 3, 4, 5, or UTP
- 100Base-FX, which uses multimode or single-mode fiber optic cable

IEEE 802.3y:

- 100Base-T2, which uses two pairs of, wires of category 3, 4 or 5 UTP

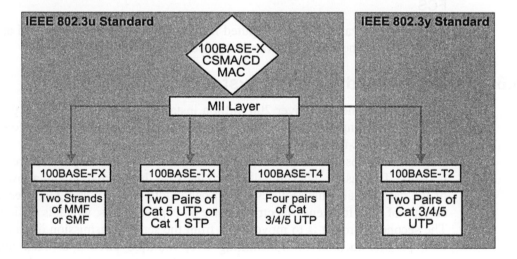

Figure 5.10
Summary of 100Base-T standards

This approach is possible because the original 802.3 specification defined the MAC layer independently of the various physical PMD layers it supports. As you will recall, the MAC layer defines the format of the Ethernet frame and defines the operation of the CSMA/CD access mechanism. The time dependent parameters are defined in the 802.3 specification in terms of bit-time intervals and so are speed independent. The 10 Mbps Ethernet InterFrameGap is actually defined as an absolute time interval of 9.60 microseconds, equivalent to 96 bit times; while the 100Mbps system reduces this by ten times to 960 nanoseconds.

One of the limitations of the 100Base-T systems is the size of the collision domain, which is 250 m. This is the maximum sized network in which collisions can be detected; being one tenth of the size of the maximum 10 Mbps network. This limits the distance between our workstation and hub to 100 m, the same as for 10Base-T, but usually only one hub is allowed in a collision domain. This means that networks larger than 200 m must be logically connected together by store and forward type devices such as bridges, routers or switches. However, this is not a bad thing, since it segregates the traffic within each collision domain, reducing the number of collisions on the network. The use of bridges and routers for traffic segregation, in this manner, is often done on industrial CSMA/CD networks.

The dominant 100Base-T system is 100Base-TX, which accounts for about 95% of all fast Ethernet shipments. The 100Base-T4 systems were developed to use four pairs of category 3 cable; however few users had the spare pairs available and T4 systems are not capable of full-duplex operation, so this system has not been widely used. The 100Base-T2 system has not been marketed at this stage, however its underlying technology using digital signal processing (DSP) techniques is used for the 1000Base-T systems on two category 5 pairs. With category 3 cable diminishing in importance, it is not expected that the 100Base-T2 systems will become significant.

5.7.2 100Base-TX and FX

This operates on two pairs of category 5 twisted pair or two multimode fibers. It uses stream cipher scrambling for data security and MLT-3 bit encoding. The multilevel threshold-3 (MLT-3) bit coding uses three voltage levels: +1 volts, 0 volts and −1 volts. The level remains the same for consecutive sequences of the same bit, i.e. continuous '1s'. When a bit changes, the voltage level changes to the next state in the circular sequence 0V, +1V, 0V, −1V, 0V etc. This results in a coded signal, which resembles a smooth sine wave of much lower frequency than the incoming bitstream. Hence for a 31.25 MHz baseband signal this allows for a 125 Mbps signaling bit stream providing a 100 Mbps throughput (4 B/5B encoder). The MAC outputs a NRZ code. This code is then passed to a scrambler, which ensures that there are no invalid groups in its NRZI output. The NRZI converted data is passed to the three level code block and the output is then sent to the transceiver. The code words are selectively chosen so the mean line signal line zero, in other words the line is DC balanced.

The three level code results in a lower frequency signal. Noise tolerance is not as high as 10Base-T because of the multilevel coding system; hence data grade (Category 5) cable is required.

Two pair wire, RJ-45 connectors and a hub are requirements for 100BASE-TX. These factors and a maximum distance of 100 m between the nodes and hubs make for a very similar architecture to 10BASE-T.

5.7.3 100BASE-T4

The 100Base-T4 systems use four pairs of category 3 UTP. It uses data encoded in an eight binary six ternary (8B/6T) coding scheme similar to the MLT-3 code. The data is encoded using three voltage levels per bit time of +V, 0 volts and −V, these are usually written as simply +, 0 and −. This coding scheme allows the eight bits of binary data to be coded into six ternary symbols and reduces the required bandwidth to 25 MHz. The 256 codewords are chosen so the line has a mean line signal of zero. This helps the receiver to discriminate the positive and negative signals relative to the average zero level. The coding utilizes only those codewords, which have a combined weight of 0 or +1, as well as at least two signal transitions for maintaining clock synchronization. For example the code word for the data byte 20H is −++−00, which has a combined weight of 0 while 40 H is −00+0+ which has a combined weight of +1.

If a continuous string of codewords of weight +1 is sent, then the mean signal will move away from zero – known as DC wander. This causes the receiver to misinterpret the data since it is assuming the average voltage it is seeing, which is now tending to '+1', is its zero reference. To avoid this situation, a string of codewords of weight +1 is always sent by inverting alternate codewords before transmission. Consider a string of consecutive data bytes 40H, the codeword is −00+0+ which has weight +1. This is sent as the sequence −00+0+, +00−0−, −00+0+, +00−0− etc, which results in a mean signal level of zero. The receiver consequently reinverts every alternate codeword before decoding.

These signals are transmitted in half-duplex over three parallel pairs of category 3, 4 or 5 UTP cable, while a fourth pair is used for reception of collision detection signals.

100BASE-TX and 100Base-T4 are designed to be interoperable at the transceivers using a media independent interface and compatible (Class 1) repeaters at the hub. Maximum node to hub distances of 100 m and 250 m maximum network diameter are supported – at a maximum hub to hub distance of 10 m.

5.7.4   100Base-T2

The IEEE published the 100Base-T2 system in 1996 as the IEEE 802.3y standard but has not been marketed at this stage. It was designed to address the shortcomings of 100Base-T4, making full-duplex 100 Mbps accessible to installations with only two category 3 cable pairs available. The standard was completed two years after 100Base-TX, at which stage the TX had such market dominance that the T2 products were not commercially produced. However it is mentioned here for reference and because its underlying technology using digital signal processing (DSP) techniques and five-level coding (PAM-5) is used for the 1000Base-T systems on two category 5 pairs.

The features of 100Base-T2 are:

- Uses two pairs of category 3, 4 or 5 UTP
- Uses both pairs for simultaneously transmitting and receiving – commonly known as dual-duplex transmission. This is achieved by using digital signal processing (DSP) techniques
- Uses a five-level coding scheme with five phase angles called pulse amplitude modulation (PAM 5) to transmit two bits per symbol.

5.7.5   100Base-T hubs

The IEEE 802.3u specification defines two classes of 100Base-T hubs, which are normally, called repeaters:

- Class I, or translational repeaters, which can support both TX/FX and T4 systems
- Class II, or transparent repeaters, which support only one signaling system

The class I repeaters have greater delays (0.7μs maximum) in supporting both signaling standards and so only permit one hub in each collision domain. The class I repeater fully decodes each incoming TX or T4 packet into its digital form at the media independent interface (MII) and then sends the packet out as an analog signal from each of the other ports in the hub. Repeaters are available with all T4 ports, all TX ports or combinations of TX and T4 ports, called translational repeaters. Their layout is shown in Figure 5.11.

The class II repeaters operate like a 10Base-T repeater connecting the ports (all of the same type) at the analog level. These then have lower inter-repeater delays (0.46 µs maximum) and so two repeaters are permitted in the same collision domain, but only 5m apart. Alternatively, in an all fiber network, the total length of all the fiber segments is 228 meters. This allows two 100 m segments to the nodes with 28 m between the repeaters and any other combination. Most fast Ethernet repeaters available today are class II.

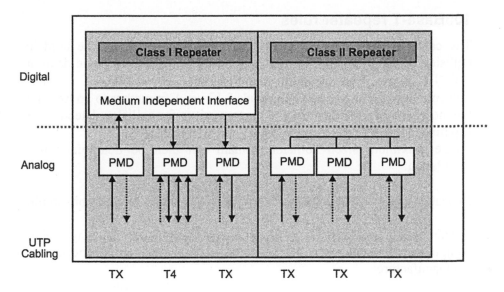

Figure 5.11
Class I and class II fast Ethernet repeaters

5.7.6   100Base-T adapters

Adapter cards are readily available as standard 100 Mbps and as 10/100 Mbps. These latter cards are interoperable at the hub on both speed systems.

5.8   Fast Ethernet design considerations

5.8.1   UTP Cabling distances 100Base-TX/T4

Maximum distance between UTP hub and desktop NIC is 100 metres, made up as follows:
- 5 meters from hub to patch panel
- 90 meters horizontal cabling from patch panel to office punch down block
- 5 meters from punch-down block to desktop NIC

5.8.2   Fiber optic cable distances 100Base-FX

The following maximum cable distances are in accordance with the 100Base-T bit budget (see Section 5.9.3)
- Node to hub: Maximum distance of multimode cable (62.5/125) is 160 meters (for connections using a single class II repeater)

- Node to switch: Maximum multimode cable distance is 210 meters
- Switch to switch: Maximum distance of multimode cable for a backbone connection between two 100Base-FX switch ports is 412 meters
- Switch to switch full-duplex: Maximum distance of multimode cable for a full-duplex connection between two 100Base-FX switch ports is 2000metres

Note: The IEEE has not included the use of single mode fiber in the 802.3u standard. However numerous vendors have products available enabling switch to switch distances of up to ten to twenty kilometers using single mode fiber.

5.8.3   100Base-T repeater rules

The cable distance and the number of repeaters, which can be used in a 100Base-T collision domain, depends on the delay in the cable and the time delay in the repeaters and NIC delays. The maximum round-trip delay for 100Base-T systems is the time to transmit 64 bytes or 512 bits and equals 5.12 µs. A frame has to go from the transmitter to the most remote node then back to the transmitter for collision detection within this round trip time. Therefore the one-way time delay will be half this.

The maximum sized collision domain can then be determined by the following calculation:

Repeater delays + Cable delays + NIC delays + Safety factor (5 bits minimum)< 2.56 µs

The following Table 5.1 gives typical maximum one-way delays for various components. Repeater and NIC delays for your specific components can be obtained from the manufacturer.

Component	Maximum delay (µs)
Fast Ethernet NIC	0.25
Fast Ethernet Switch Port	0.25
Class I Repeater	0.7 max
Class II Repeater	0.46 max
UTP Cable (per 100 meters)	0.55
Multimode Fiber (per 100 meters)	0.50

Table 5.1
Maximum one-way fast Ethernet component delays

Notes: If the desired distance is too great it is possible to create a new collision domain by using a switch instead of a repeater. Most 100Base-T repeaters are stackable, which means multiple units can be placed on top of one another and connected together by means of a fast backplane bus. Such connections do not count as a repeater hop and make the ensemble function as a single repeater.

5.9   Gigabit Ethernet 1000Base-T

5.9.1   Gigabit Ethernet summary

Gigabit Ethernet uses the same 802.3 frame format as 10 Mbps and 100 Mbps Ethernet systems. This operates at ten times the clock speed of fast Ethernet at 1 Gbps. By retaining the same frame format as the earlier versions of Ethernet, backward compatibility is assured with earlier versions, increasing its attractiveness by offering a high bandwidth connectivity system to the Ethernet family of devices.

Gigabit Ethernet is defined by the IEEE 802.3z standard. This defines the gigabit Ethernet media access control (MAC) layer functionality as well as three different physical layers: 1000Base-LX and 1000Base-SX using fiber and 1000Base-CX using copper. IBM originally developed these physical layers for the ANSI fiber channel systems and used 8B/10B encoding to reduce the bandwidth required to send high speed signals. The IEEE merged the fiber channel to the Ethernet MAC using a gigabit media independent interface (GMII) which defines an electrical interface enabling existing fiber channel PHY chips to be used and enabling future physical layers to be easily added.

1000Base-T is being developed to provide service over four pairs of category 5 or better copper cable. As discussed earlier this uses the same technology as 100Base-T2. This development is defined by the IEEE 802.3ab standard.

These gigabit Ethernet versions are summarized in Figure 5.12.

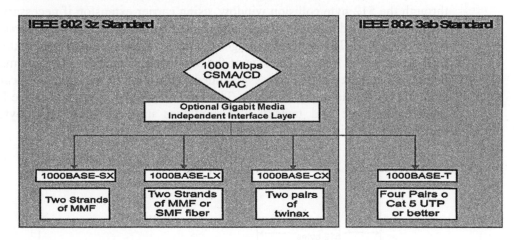

Figure 5.12
Gigabit Ethernet versions

5.9.2   Gigabit Ethernet MAC layer

Gigabit Ethernet retains the standard 802.3 frame format, however the CSMA/CD algorithm has had to undergo a small change to enable it to function effectively at 1 Gbps. The slot time of 64 bytes, used with both 10 Mbps and 100 Mbps systems, has been increased to 512 bytes. Without this increased slot time, the network would have been impractically small at one tenth of the size of fast Ethernet – only 20 meters.

The slot time defines the time during which the transmitting node retains control of the medium, and in particular is responsible for collision detection. With gigabit Ethernet it was necessary to increase this time by a factor of eight to 4.096 μs to compensate for the tenfold speed increase. This then gives a collision domain of about 200 m.

If the transmitted frame is less than 512 bytes the transmitter continues transmitting to fill the 512 byte window. A carrier extension symbol is used to mark frames, which are shorter than 512 bytes and to fill the remainder of the frame. This is shown in Figure 5.13.

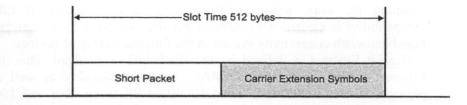

Figure 5.13
Carrier extension

While this is a simple technique to overcome the network size problem, it could cause problems with very low utilization if we send many short frames, typical of some industrial control systems. For example, a 64 byte frame would have 448 carrier extension symbols attached and result in a utilization of less than 10%. This is unavoidable, but its effect can be minimized if we are sending many small frames by a technique called packet bursting. Once the first frame in a burst has successfully passed through the 512 byte collision window, using carrier extension if necessary, transmission continues with additional frames being added to the burst until the burst limit of 1500 bytes is reached. This process averages the time wasted sending carrier extension symbols over a number of frames. The size of the burst varies depending on how many frames are being sent and their size. Frames are added to the burst in real-time with carrier extension symbols filling the interpacket gap. The total number of bytes sent in the burst is totaled after each frame and transmission continues until at least 1500 bytes have been transmitted. This is shown in Figure 5.14.

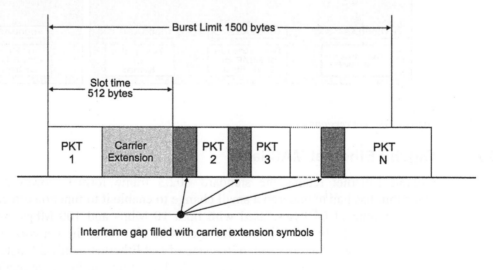

Figure 5.14
Packet bursting

5.9.3   1000Base-SX for horizontal fiber

This gigabit Ethernet version was developed for the short backbone connections of the horizontal network wiring. The SX systems operate full duplex with multimode fiber only, using the cheaper 850 nm wavelength laser diodes. The maximum distance supported varies between 200 and 550 meters depending on the bandwidth and attenuation of the fiber optic cable used. The standard 1000Base-SX NICs available today are full-duplex and incorporate SC fiber connectors.

5.9.4   1000Base-LX for vertical backbone cabling

This version was developed for use in the longer backbone connections of the vertical network wiring. The LX systems can use single mode or multimode fiber with the more expensive 1300 nm laser diodes. The maximum distances recommended by the IEEE for these systems operating in full-duplex is 5 kilometers for single mode cable and 550 meters for multimode fiber cable. Many 1000Base-LX vendors guarantee their products over much greater distances, typically 10 km. Fiber extenders are available to give service over as much as 80 km. The standard 1000Base-LX NICs available today are full-duplex and incorporate SC fiber connectors

5.9.5   1000Base-CX for copper cabling

This version of gigabit Ethernet was developed for the short interconnection of switches, hubs or routers within a wiring closet. It is designed for 150 ohm twinax cable similar to that used for IBM token ring systems. The IEEE specified two types of connectors: The high-speed serial data connector (HSSDC) known as the fiber channel style 2 connector and also the nine pin D-subminiature connector from the IBM token ring systems. The maximum cable length is 25 meters for both full- and half-duplex systems.

These systems are not currently available in the marketplace for connecting different switches. The preferred connection arrangements are to connect chassis-based products via the common backplane and stackable hubs via a regular fiber port.

5.9.6   1000Base-T for category 5 UTP

This version of the gigabit Ethernet is developed under the IEEE 802.3ab standard for transmission over four pairs of category 5 or better cable. This is achieved by simultaneously sending and receiving over each of the four pairs. Compare this to the existing 100Base-TX system, which has individual pairs for transmitting and receiving.

This system uses the same data encoding scheme developed for 100Base-T2, which is PAM5. This utilizes five voltage levels so has less noise immunity, however the digital signal processors (DSP) associated with each pair overcomes any problems in this area. The system achieves its tenfold speed improvement over 100Base-T2 by transmitting on twice as many pairs (4) and operating at five times the clock frequency (125 MHz).

5.9.7   Gigabit Ethernet full-duplex repeaters

Gigabit Ethernet nodes are connected to full-duplex repeaters also known as non-buffered switches or buffered distributors. These devices have a basic MAC function in each port, which enables them to verify that a complete frame is received and compute its frame check sequence (CRC) to verify the frame validity. Then the frame is buffered in the internal memory of the port before being forwarded to the other ports of the repeater. It is therefore combining the functions of a repeater with some features of a switch.

All ports on the repeater operate at the same speed of 1 Gbps, and operate in full duplex so it can simultaneously send and receive from any port. The repeater uses 802.3x flow control to ensure the small internal buffers associated with each port do not overflow. When the buffers are filled to a critical level, the repeater tells the transmitting node to stop sending until the buffers have been sufficiently emptied. The repeater does not analyze the packet address fields to determine where to send the packet, like a switch does, but simply sends out all valid packets to all the other ports on the repeater.

The IEEE does allow for half-duplex gigabit repeaters – however none exist at this time.

5.10 Network interconnection components

Distances of LANs are often limited and there is often a need to increase this range. There is a number of interconnecting devices, which can be used to achieve this ranging from repeaters to routers to gateways. It may also be necessary to partition an existing network into separate networks for reasons of security or traffic overload.

These components to be discussed separately are:

- Repeaters
- Bridges
- Routers
- Gateways
- Hubs
- Switches

5.10.1 Repeaters

A repeater operates at the physical layer of the OSI model and simply retransmits an incoming electrical signal. This means simply amplifying and retiming the signal received on one segment onto all other segments. All segments need to operate with the same media access mechanism and the repeater is unconcerned with the meaning of the individual bits in the data. Collisions, truncated packets or electrical noise on one segment are transmitted onto all other segments.

The main reason for the use of repeaters is to extend the segment beyond the recommended length.

The number of repeaters is generally restricted to two (but some recommend a maximum of four). Timing problems occur when too many repeaters are used.

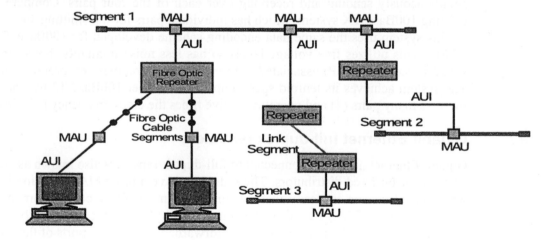

Figure 5.15
Ethernet repeater

Segments connected by repeaters should generally have similar traffic, since all traffic is repeated to the other segments.

Another variation on the standard repeater is the multi-port repeater, which connects more than two segments. A useful application of multi-port repeaters is connecting different cable media together (e.g. thick coaxial to thin coaxial or twisted pair). Multi-port repeaters are sometimes also referred to as multimedia concentrators.

5.10.2   Bridges

Bridges are used to connect two separate networks to form a logical network. The bridge has a node on each network and passes only valid messages across to destination addresses on the other network(s). The bridge stores the frame from one network and examines its destination address to determine whether it should be forwarded over the bridge. Figure 5.16 shows the basic configuration of an Ethernet Bridge.

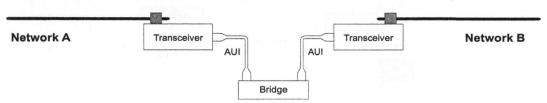

Figure 5.16
Ethernet bridge

The bridge maintains records of the Ethernet addresses of the nodes on both networks to which it is connected. The data link protocol must be identical on both sides of the bridge, however, the physical layers (or cable media) do not necessarily have to be the same. Thus, the bridge isolates the media access mechanisms of the networks. Data can therefore be transferred between Ethernet and token ring LANs. For example, collisions on the Ethernet system do not cross the bridge nor do the tokens. The bridge provides a transparent connection between a full size LAN with maximum count of stations, repeaters and cable lengths and any other LAN.

Bridges can be used to extend the length of a network (as with repeaters) but in addition, they improve network performance. For example, if a network is demonstrating slow response times, the nodes that mainly communicate with each other can be grouped together on one segment and the remaining nodes can be grouped together in another segment. The busy segment may not see much improvement in response rates (as it is already quite busy) but the lower activity segment may see quite an improvement in response times. Bridges should be designed so that 80% or more of the traffic is within the LAN and only 20% crosses the bridge. Stations generating excessive traffic should be identified by a protocol analyzer and relocated to another LAN.

5.10.3   Router

Routers are used to transfer data between two networks that have the same network layer protocols (such as TCP/IP) but not necessarily the same physical or data link protocols. Figure 5.17 shows the application of routers.

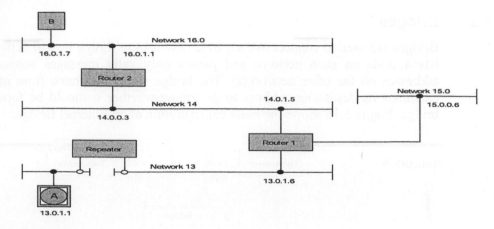

Figure 5.17
Router applications

The routers maintain tables of the networks to which they are attached and to which they can route messages. Routers use the network (IP) address to determine where the message should be sent, because the network address contains routing information. Routers maintain tables of the optimum path to reach particular network and redirect the message to the next router along that path.

For example in Figure 10.3, to transmit a message from node A on network 13.0 to node B on network 16.0, router 1 forwards the message onto network 14.0 and router 2 forwards it to network 16.0.

Router 1 stores the message from A and examines the IP address of the destination (16.0.1.7). Consulting its routing table it determines that the message needs to be sent to router 2 in order to reach network 16.0. It then replaces the destination hardware address on the message with router 2's address and forwards the message on network 14.0. Router 2 repeats the process and determines from the destination IP address (16.0.1.7) that it can deliver the message directly on network 16.0. Router 2 establishes the hardware address (16.0.1.7) from its routing table and places that destination address on the message which it delivers on network 16.0.

5.10.4   Gateways

A gateway is designed to connect dissimilar networks. A gateway may be required to decode and re-encode all seven layers of two dissimilar networks connected to either side. Gateways thus have the highest overhead and the lowest performance of the internetworking devices. For example, a gateway could connect an Ethernet network and a token ring network. The gateway translates from one protocol to the other (possibly all seven layers of the OSI model) and handles difference in physical signals, data format and speed.

5.10.5   Hubs

Hubs are used to implement physical star networks for 10BaseT and token ring systems in such a way that electrical problems on individual node-to-hub links would not affect the entire network.

Hubs generally are of two types – cabinet hubs and chassis hubs. The former is a single sealed cabinet with all the connections attached and no internal expansion capability. These units are of low initial cost and are small.

Chassis hubs provide a cabinet with a backplane for interconnecting cable connection modules. Although more expensive these chassis hubs provide greater flexibility, improve network reliability by removing vulnerable interhub cabling and allow expansion across the hub backplane without needing additional repeaters.

5.10.6 Switches

One of the devices that has made a major change to the way networks and Ethernet are used today is the ubiquitous switch. This enables direct communications between multiple pairs of devices in full duplex mode; thus eliminating the limitations imposed by the classical Ethernet architecture.

Switches enable specific and discrete data transfers to be accomplished between any pair of devices on a network, in a dedicated fashion. Stemming from the STAR network, where each terminal had its own discrete cable to a central hub, it became apparent that there was a requirement for a network to be able to quickly and effectively connect 2 terminals or nodes together. This is to be done in such a manner that they had, in the case of Ethernet 10baseT a direct and dedicated 10 Mbps connection.

Illustrated below is an example of a Cisco Kalpana 8 port switch, with 8 Ethernet terminals attached. This comprises a star configuration, but it operates in a different manner.

Nodes or terminals 1 & 7, 3 & 5 and 4 & 8 are connected directly. For example, terminal 7, detecting CSMA/CD on its local connection sends a packet to the switch. The switch determines the destination address and directs the package to the corresponding port; Port 1 in this example. Data is then passed at 10 Mbps between the two devices.

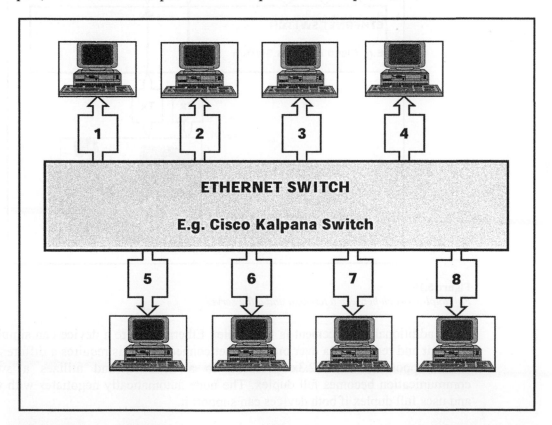

Figure 5.18
Ethernet switch

If terminal 3 wishes to communicate with terminal 5, the same procedure is initiated by device 3 as it was for terminal 7 as described above. In fact, if the switch was say a 16 port switch (an 8 port being shown in this example) it would be possible for 8 pairs of terminals to be simultaneously communicating between themselves, all at the Ethernet rate of 10 Mbps for each established circuit.

The switch must be able to detect, from the packet header, the Ethernet destination address, and effect the required port connection in time for the remainder of the packet to be switched through. Some switches cannot achieve this requirement, their switching is slower than the effective requirement made on them by the transmission speeds. These switches are known as store-and-forward switches, and delay the packet by the time required to store, switch and forward it.

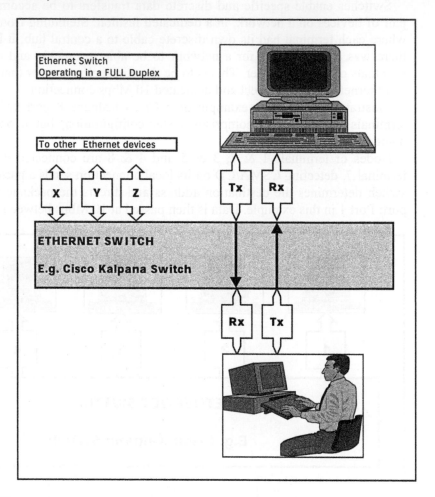

Figure 5.19
Full duplex switch between a terminal and a file server

An additional advancement is full duplex Ethernet where a device can simultaneously transmit and receive data over one Ethernet connection. This requires a different Ethernet card, supporting the 802.3x protocol, to a terminal, and utilizes a switch; the communication becomes full duplex. The node automatically negotiates with the switch and uses full duplex if both devices can support it.

This form of configuration is useful in places where large amounts of data require to be moved around quickly. For example, graphical workstations, large color plotters and fileservers.

5.11 TCP/IP protocols

The TCP/IP layering scheme organizes the protocol software into four layers residing on top of the hardware layer. The ISO model can be modified to accommodate the TCP/IP model, but there are major differences between them. Communication between two different nodes is as indicated in Figure 5.20.

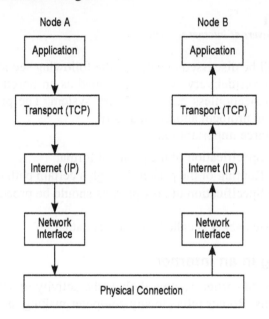

Figure 5.20
Communication between two different nodes using TCP/IP

- Application layer: At the highest level the application programs, interact with the transport level protocol to send or receive data.
- Transport layer: The transport layer provides end-to-end communications. It ensures that data arrives without errors and in the correct sequence.
- Internet layer: This performs the encapsulation of the request from the transport layer in an IP datagram, uses the routing algorithms to determine whether to deliver the datagram directly or send it to a gateway. The Internet layer sends ICMP error and control messages as needed.
- Network interface or physical layer: This is responsible for accepting IP datagrams and transmitting them over a specific network.
- Protocol layering: The layering of protocols ensures that programs can be written in a modular fashion. A particular layer at one node will communicate directly with a layer at the other node.

5.11.1   The TCP/IP protocol structure

TCP/IP provides three layers of services as indicated in Figure 5.21.

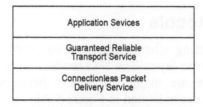

Figure 5.21
The three layers of Internet services

These will be discussed in detail in the following sections.

The packet delivery system is defined as an unreliable (no guaranteed delivery), best effort, connectionless packet delivery system. The protocol that describes this is called the Internet protocol abbreviated as IP.

IP has three important functions:

- Specification of the protocol format
- Routing of the packet through a certain path of the Internet
- Specification of how packets should be processed and how to handle errors, etc.

The basic packet is called an Internet datagram.

5.11.2   Routing in an Internet

As discussed earlier, routing defines the activity of selecting the path over which to send the packets. Router refers to any computer making the decision on which path or route to use.

Both nodes and gateways are involved in the routing of the IP message.

There are two forms of routing:

- Direct routing where a node transfers a message directly to another, all connected on the same network
- Indirect routing where the destination node is not directly connected and the message has to be passed to a gateway for delivery

As discussed earlier, for direct routing on a particular physical network, the transmitting node encapsulates the datagram in a physical frame (such as Ethernet), binds the destination Internet address to the physical hardware address and physically transmits the message. The transmitter easily knows whether the transfer problem is direct routing as it compares the network portion of the destination IP address with its own network identification. This is quickly done within the software.

Indirect routing is done by software algorithms (and data tables) in each gateway examining the datagram's destination network address (as it arrives) and then forwarding the datagram to the next appropriate gateway (in the direction of the destination node). Note that the gateway routing software is only concerned with the network address – the node address is not relevant.

The technique used for efficient routing of the datagram is to use an Internet routing table, which stores information on all destination networks and how to reach them. The routing table can only point to gateways that are physically connected to the network to which the gateway (which currently has the datagram) is connected. The size of the routing table depends on the number of interconnected networks. The number of individual nodes connected to the various networks does not affect the size of the routing table. An example is given in the figure below for routing from G$_B$.

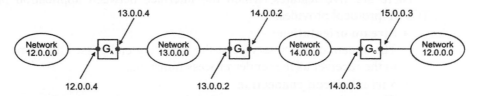

To reach nodes on network	Route to this address
12.0.0.0	13.0.0.4
13.0.0.0	Deliver Directly
14.0.0.0	Deliver Directly
15.0.0.0	14.0.0.3

Figure 5.22
Example of routing table for G$_B$

When the routing algorithm extracts the next hop address (where the datagram must be sent to next in its path), it passes this information and the datagram to the physical network software, which binds the next hop address to a physical address. The physical network software then forms a frame using that physical address as the destination address (for this one hop only), and places that datagram in the data section of the message and dispatches the packet. The reason for the IP routing software not using or calculating the physical addresses is to isolate the IP software from the physical details of the network (and hide the underlying details of the network from the IP software).

When the datagram arrives at a destination host (or node), the physical network interface software transfers it to the IP software for further processing. There are two cases:

- The datagram IP destination address matches that of the destination node and the message is processed by the IP software
- The datagram destination IP address does not match and the node is then required to get rid of the datagram (and not to forward it to another node)

5.11.3   Transmission control protocol (TCP)

Although this is generally considered part of the TCP/IP Internet protocol suite, it is a protocol that can be used independently. It can be considered to operate independently of the underlying physical network.

The transmission control protocol (TCP) specifies the structure of the messages, the acknowledgments between two nodes for reliable data transfer, how messages are routed to multiple destinations on a machine, and how errors are detected and corrected.

TCP is quite efficient although it is written for a general application. It can run at 8 Mbps for two workstations operating on a 10 Mbps or with a super computer operate at up to 600 Mbps.

There is a need for a reliable means of delivering message packets. Packets can be lost or destroyed with transmission errors, or when networks become too heavily loaded to cope with an overloaded situation. Application programs at the highest level need to be able to send data from one point to another in a reliable manner. It is important to have a general-purpose protocol such as TCP to handle the problems of reliable transmission of data, which the application programs can then use.

There are five features, which the interface between application programs and the TCP/IP protocol provides:

- **Stream orientation**
  The stream delivery service at the receiving end passes the same sequence of bytes to the receiver as the sender process transmits.
- **Virtual circuit connection**
  The application programs can view the connection (between the source and destination nodes) as a virtual circuit, when using the TCP interface. The TCP software establishes a connection between the two systems and then advises the application programs that transmission of data can proceed.
- **Buffered transfer**
  The protocol software takes the data stream from the application program and breaks it up into an appropriate size for transmission across the communication links. If the data has to be broken up into smaller components for transfer across the link, there will be a need for buffering of the data.
- **Unstructured stream**
  The TCP/IP protocol cannot handle structured data streams with boundaries marked (in the data stream) to indicate different types of data. It is the responsibility of the application program to agree on the format of the data stream.
- **Full duplex connection**
  This allows transfer of data in both directions at the same time. Piggy-backup is a useful feature where control instructions can be embedded in the datagrams carrying data back to the source.

5.12   SCADA and the Internet

The Internet is simply the single (virtual) network where all stations are easily connected together without worrying about the underlying physical connections. In connecting two nodes together, the communications path may be across multiple networks (or local area networks) to which neither node is connected. This is made possible by the use of a universal family of open protocols called TCP/IP, which form the basis of the Internet.

Essentially the IP protocol offers the ability to perform routing allowing packets to be sent over a rather complicated topology of interconnected networks. The TCP portion of the protocol allows the packets to be sent from one point to another and to have a cast-iron guarantee that they do indeed arrive at the required destination

The application layer protocols can be items such as the hypertext transfer protocol (HTTP) used by the World Wide Web (www). The www, or simply web, is the graphical interface that enables you to read and download information people have stored in the standard www format. Although the Internet is extremely popular and a great way of communicating between stations, the intranet is another term, which is becoming important in the SCADA world. This is simply put – an internal Internet network peculiar to a specific company and not necessarily connected to the Internet.

A comparison between the Internet, intranet and a standard local area network is given in Figure 5.23 below.

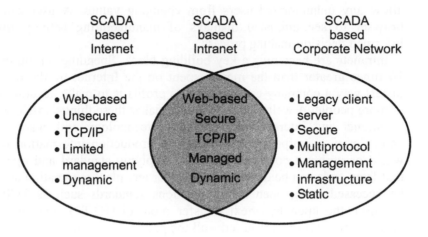

SCADA based Internet   SCADA based Intranet   SCADA based Corporate Network

- Web-based
- Unsecure
- TCP/IP
- Limited management
- Dynamic

Web-based Secure TCP/IP Managed Dynamic

- Legacy client server
- Secure
- Multiprotocol
- Management infrastructure
- Static

Figure 5.23
Comparison of Internet/Intranet and (legacy) corporate LAN
(Courtesy of 3Comm)

An intranet is designed to communicate within a single network. Intranets can be defined as IP-based corporate networks that employ a standard web browser as the standard user interface for all workstations connected to the network. Intranets offer a superb way of capitalizing on the benefits of the Internet technologies – a consistent interface throughout the network, easy-to-use web publishing tools and languages, mixed media, centralized maintenance of shared resources and more – all in an internal network. Many companies with a SCADA system may want to restrict the data transfer to the company environment only.

5.12.1   Use of the Internet for SCADA systems

With the addition of a modem and some software, these factory-floor PC hosts become Internet nodes accessible from anywhere in the world. Clients and system integrators can log into a remote factory and can do anything they could do if they were in the same building. They can look at the revision number of the CPU and observe the program running. Since local technicians can also log onto the network, they can easily collaborate with more sophisticated support back at some world-wide support center.

Hence the Internet has the potential to render many traditional SCADA, telemetry and data acquisition systems with their expensive RF, dial-up and leased-line communications obsolete.

With readily available software and hardware one can put together a data acquisition system which can capture real-time data and transmit it anywhere in the world today at a negligible cost.

5.12.2   Thin client solutions

There are a number of SCADA software companies offering solutions in this area. A good example is the FIX web server from Intellution. This is a 'thin client' solution that enables plant, management, production, and maintenance personnel to view real-time process graphics from a remote location using a standard Internet or intranet connection. This provides read-only access to real-time graphics and industrial automation software from the World Wide Web and/or intranet. Central SCADA package's graphics can be viewed using any standard Internet browser. However, it protects data because it does not allow any unauthorized users from changing values. A user can use a standard web browser to see animated displays of manufacturing activity thus allowing a more informed decision-making process.

Intranets are becoming a key building block. Spending on intranets is projected to be 10 times greater than the money spent on the Internet by the turn of the century. The fastest way a company can improve its profits is to make accurate information available to more people. A well-planned intranet makes these things possible.

Intranet and Internet communications technology will change the user's ability to gather or view corporate information or production information from any place in the world or within a corporate network. Applying standard and open technologies means that front ends can be developed easily for any plant automation application and they can be accessed using open communications standards such as TCP/IP and Ethernet. In addition, the open hypertext transfer protocol (HTTP) runs on top of TCP/IP so no proprietary network is required – all the protocols are open.

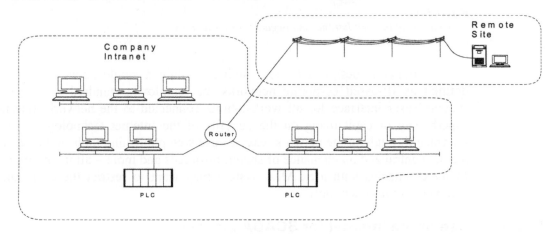

Figure 5.24
Intranets and the Internet

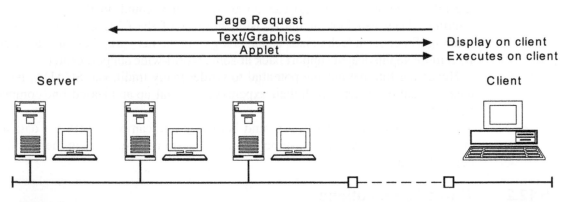

Figure 5.25
Operation of the web

5.12.3   Security concerns

Areas of caution can be summed up with the word security. At the time of writing, there have not been many incidents, which would make the plant manager concerned about his

SCADA system using the Internet technologies. However, the problems in the commercial world with breakdowns in security will definitely manifest themselves in the SCADA arena as well. Hence, new safeguards need to be introduced to protect sensitive information and resources.

In the practical world, most users will not want to actually run a plant remotely via the Internet but will prefer to use the Internet mainly as a remote access facility to view real-time data at the site. One would not want multiple people at different parts of the world making changes to critical parameters on a plant.

The overall communications system must be dynamic allowing for rapid shifts in the traffic pattern of the SCADA system whilst still delivering a specified level of performance. While the TCP/IP protocols may be widely used, they contribute a considerable amount of overhead to a packet sent from one station to another. Hence, if the amount of information being sent is only a few bytes; most of the transmission time will be occupied with overhead bytes, which make no direct contribution to the information transfer. The Ethernet networks used by the Internet system do not guarantee speed. Hence, they have to be sized correctly to ensure in the worst case of a high level of traffic (e.g. plant shutdown) that the messages do indeed get through in the required time.

5.12.4   Other issues

A few other issues are listed below:

- The SCADA system will require universal IP access across the network. This will allow anyone at numerous locations to be able to access the SCADA system.
- The overall SCADA system (based on either the Internet or intranet) must be manageable. This means the system administrator needs to have clear information about the data flows and traffic across all points of the system.
- Finally the overall system must be adaptable; able to cope with the changing requirements of the SCADA system and the business which it is serving.

5.12.5   Conclusion

Technical excellence (naturally) is admired as a desirable attribute of any control system. However, the driving force today will be the rapid reduction of costs. The Internet is a key element in reduction of costs, providing us with open systems freely available to all and allowing the rapid communication of information to all stakeholders in an enterprise. The typical SCADA system will increasingly be built upon the Internet protocols. This will increasingly allow the SCADA vendor to focus on providing excellent application software whilst relying on the open Internet protocols to connect the different components of the network together.

6

Modems

6.1 Introduction

This chapter reviews the concept of and practical use of a modem in a telemetry system. The following topics will be examined in this section:

- Review of the modem
- The RS-232/RS-422/RS-485 standards
- Flow control techniques
- Modulation techniques
- Error detection/correction and data compression
- Data rate versus baud rate
- Modem standards
- Radio modem systems
- Troubleshooting modem systems
- Selection considerations for modems

6.2 Review of the modem

The telephone system, landline communication systems and radio systems cannot directly transport digital information without some distortion in the signal due to the bandwidth limitation inherent in the connecting medium. The reason for the difficulty in transferring digital information over a telephone network, for example, is the limited bandwidth inherent in the communication media, such as telephone cable with capacitance and inductance. The bandwidth (defined as the difference between the upper and lower allowable frequency) is typically 300 Hz to 3400 Hz for telephone cable. This is illustrated in Figure 6.3.

A conversion device, called a modem (modulator/demodulator), is thus required to convert the digital signals into an analog form suitable for transmission over a telephone network. This converts the digital signals generated by a computer into an analog form suitable for long distance transmission over the cable or radio system. The demodulation portion of the modem receives this analog information and converts it back into the original digital information generated by the transmitting computer.

Figure 6.1 gives a schematic view of the modem's place in the communications hierarchy.

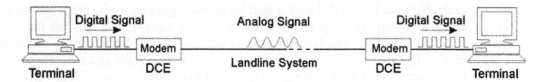

Figure 6.1
The modem as a component in an overall system

An example of what a digital signal would look like at the far end of a cable if it were injected directly into the cable is given in Figure 6.2.

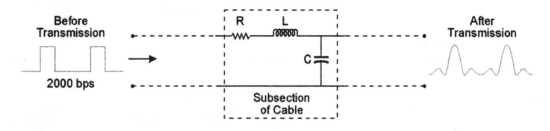

Figure 6.2
Injection of a digital signal down a cable

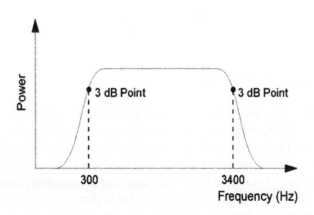

Figure 6.3
The bandwidth limitation problem

There are two types of modem available today:
- Dumb (or non-intelligent) modems depend on the computer to which they are connected, to instruct the modem when to perform most of the tasks such as answering the telephone.
- Smart modems have an on-board microprocessor enabling them to perform such functions as automatic dialing and the method of modulation to use.

6.2.1 Synchronous or asynchronous

Modems can be either synchronous or asynchronous. In asynchronous communications each character is encoded with a start bit at the beginning of the character bit stream and a parity and stop bit at the end of the character bit stream. The receiver then synchronizes with each character received by looking out for the start bit. Once the character has been received, the communications link returns to the idle state and the receiver watches out for the next start bit (indicating the arrival of the next character). This will be discussed in more detail under the RS-232 standard in Section 6.3.1.

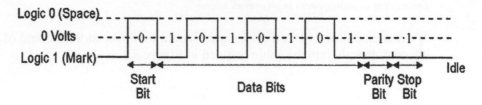

Figure 6.4
Format of a typical serial asynchronous data message

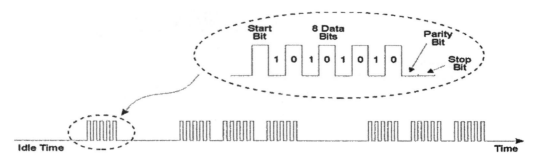

Figure 6.5
Asynchronous transmission of a few characters

Synchronous communication relies on all characters being sent in a continuous bit stream. The first few bytes in the message contain synchronization data allowing the receiver to synchronize onto the incoming bit stream. Hereafter synchronization is maintained by a timing signal or clock. The receiver follows the incoming bit stream and maintains a close synchronization between the transmitter clock and receiver clock. Synchronous communications provides for far higher speeds of transmission of data, but is avoided in many systems because of the greater technical complexity of the communications hardware.

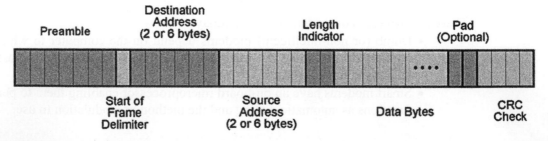

Figure 6.6
Synchronous communications protocol

6.2.2   Modes of operation

Modems can operate in two modes:
- Half-duplex
- Full-duplex

The sketches below give an example of each mode.

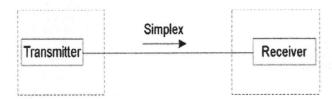

Figure 6.7
Diagram of simplex communication

A simplex system in data communications is one that is designed for sending messages in one direction only and has no provision for sending data in the reverse direction.

A duplex system in data communications is one that is designed for sending messages in both directions. Duplex systems are said to be half-duplex when messages and data can flow in both directions but only in one direction at a time.

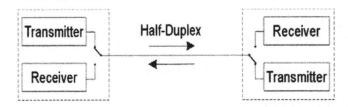

Figure 6.8
Diagram of half-duplex communication

Duplex systems are said to be full-duplex when messages can flow in both directions simultaneously.

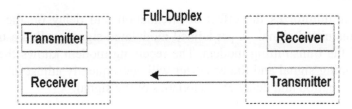

Figure 6.9
Diagram of full-duplex communication

Full-duplex is more efficient, but requires a communication capacity of at least twice that of half-duplex.

6.2.3 Components of a modem

The various components of a modem are indicated in Figure 6.10.

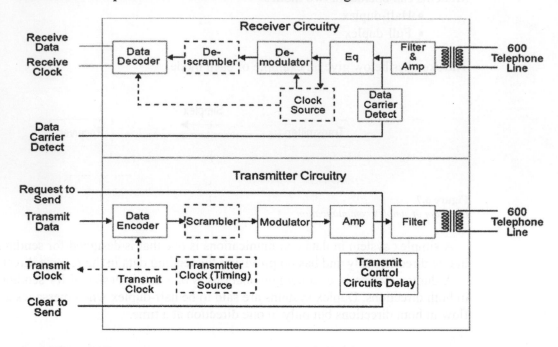

Figure 6.10
Basic components of a modem

The basic components of a modem are:

6.2.4 Modem receiver

Filter and amplifier

Noise is removed from the signal and the resultant signal is then amplified.

Equalizer

This minimizes the effect of attenuation and delay on the various components of the transmitted signal. A predefined modulated signal (called a training signal) is sent down by the transmitting modem. The receiving modem knows the ideal characteristics of the training signal and the equalizer will consequently adjust its parameters to correct for the attenuation and delay characteristics of the signal.

Demodulator

This retrieves the bit stream from the analog signal.

Descrambler (synchronous operation only)

This restores the data to its original serial form after it has been encoded in the scrambler circuit to ensure that long sequences of 1's and 0's do not occur. Long sequences of 1's and 0's are difficult to use in synchronous circuits because of the difficulty of extracting clocking information.

Data decoder

The final bit stream is produced here in true RS-232 format.

6.2.5 Modem transmitter

Data encoder

This takes the serial bit stream and uses multilevel encoding (where each signal change represents more than one bit of data) to encode the data. Depending on the modulation technique used the bit rate can be two or four (or more) times the baud rate.

Scrambler (synchronous operation only)

The bit stream is modified so that long sequences of 1's and 0's do not occur (with consequent problems for the receiver not being able to extract the clock rate).

Modulator

The bit stream is converted into the appropriate analog form using the selected modulation technique. Where initial contact is established with the receiving modem, a carrier is put on the line initially.

Amplifier

This increases the level of the signal to the appropriate level for the telephone line and matches the impedance of the line.

There are two significant causes of distortion in the signal (along a landline for example). These are 'attenuation distortion' and 'envelope delay' both indicated in Figure 6.11.

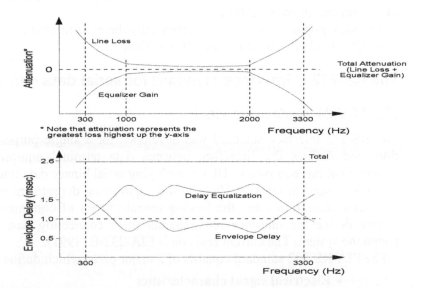

Figure 6.11
Attenuation distortion and envelope delay

Attenuation distortion indicates that the smooth (horizontal) plot of power transmitted versus frequency is not realized in practice; but that the higher frequencies tend to attenuate more easily and of course attenuation becomes notably more non-linear at the edges of the passband. Hence, the equalizer compensates with an equal and opposite effect thus giving a constant total loss throughout the operating bandwidth (or passband).

Delay distortion reflects the reality of transmission of signals down a line where the phase change to frequency is non-linear i.e., the phase tends to alter as the signal is transmitted down the communications link. The phase delay is calculated by dividing the phase by the frequency of the signal at any point along the line. The slope of phase versus frequency is called the envelope delay. Delay distortion causes problems in that two different frequencies (indicating a '1' or a '0' bit) interfere with each other at the receiving modem thus causing a potential error (called intersymbol interference).

6.3 The RS-232/RS-422/RS-485 interface standards

RS-232, RS-422 and RS-485 form the key element in transferring digital information between the RTUs (or operator terminals), and the modems, which convert the digital information to the appropriate analog, form suitable for transmission over greater distances.

These data communications standards will be examined in the detail necessary for a complete understanding of the telemetry sections of the workshop.

An interface standard defines the electrical and mechanical details that allow communications equipment from different manufacturers to be connected together and to function efficiently. It should be emphasized that RS-232, and the other related EIA standards, define the electrical and mechanical details of the interface and do not define a protocol.

These standards were designed primarily to transport digital data from one point to another. The RS-232 standard was initially designed to connect digital computer equipment to a modem where the data would then be converted into an analog form suitable for transmission over greater distances. The RS-422 and RS-485 standards can perform the same function but also have the ability of being able to transfer digital data over distances of over 1200 m.

The most popular (but probably technically the most inferior) of the RS standards is the RS-232C standard. This will be discussed first.

6.3.1 The RS-232-C interface standard for serial data communication

(CCITT V.24 interface standard)

The RS-232 interface standard was developed for a single purpose clearly stated in its title, and defines the 'interface between data terminal equipment (DTE) and data communications equipment (DCE) employing serial binary data interchange'.

It was issued in USA in 1969 by the engineering department of EIA, in cooperation with Bell laboratories and the leading manufacturers of communications equipment, to clearly define the interface requirements when connecting data terminals to the Bell telephone system. The current revision is EIA-232-E (1991).

The EIA-RS-232 standard consists of 3 major parts, which define:

- **Electrical signal characteristics**
  Electrical signals such as the voltage levels and grounding characteristics of the interchange signals and associated circuitry.

- **Interface mechanical characteristics**
  This section defines the mechanical characteristics of the interface between DTE and DCE. It dictates that the interface must consist of a 'plug' and

'receptacle' (socket) and that the receptacle will be on the DCE. In RS-232-C, the pin number assignments are specified but, originally, the type of connector was not.

- **Functional description of the interchange circuits**
  This section defines the function of the data, timing, and control signals used at the interface between DTE and DCE.

6.3.2   Electrical signal characteristics

The RS-232 interface standard is designed for the connection of two devices called:

- **DTE – data terminal equipment (e.g. a computer or printer)**
  A DTE device communicates with a DCE device. A DTE device transmits data on pin 2 and receives data on pin 3.

- **DCE – data communications equipment**
  Also called data circuit-terminating equipment in RS-232-D/E (e.g. a computer or modem). A DCE device transmits data between the DTE and a physical data communications link (e.g. telephone system). A DCE device transmits data on pin 3 and receives data on pin 2.

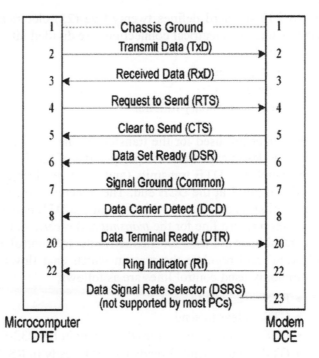

Figure 6.12
The connections between the DTE and DCE

At the RS-232 receiver, the following signal voltage levels are specified:

	Voltage Between		
Logic 0	+3V	and	+25V
Logic 1	–3V	and	–25V
Undefined	–3V	and	+ 3V

Table 6.1
RS-232 receiver voltages

Similarly, the RS-232 transmitter has to produce a slightly higher voltage level in the range of +5 volts to +25 volts and –5 volts to –25 volts to overcome the voltage drop along the line. In practice, most transmitters operate at voltages between 5 volts and 12 volts.

The RS-232 standard defines twenty-five (25) electrical connections, which are each described later. The electrical connections are divided into the four groups shown below:

- Data lines
- Control lines
- Timing lines
- Special secondary functions

The data lines are used for the transfer of data. Pins 2 and 3 are used for this purpose. Data flow is designated from the perspective of the DTE interface. Hence the 'transmit line', on which the DTE transmits (and DCE receives), is associated with pin 2 at the DTE end and pin 2 at the DCE end. The 'receive line', on which the DTE receives (and DCE transmits), is associated with pin 3 at the DTE end and pin 3 at the DCE end. Pin 7 is the common return line for the *transmit* and *receive* data lines.

The control lines are used for interactive device control, commonly known as 'hardware handshaking' and regulate the way in which data flows across the interface. The four most commonly used control lines are as follows:

- RTS – request to send
- CTS – clear to send
- DSR – data set ready (or DCE ready in RS-232-D/E)
- DTR – data terminal ready (or DTE ready in RS-232-D/E)

Note that the handshake lines operate in the opposite voltage sense to the data lines. When a control line is active (logic=1), the voltage is in the range +3 to +25 volts and when deactivated (logic=0), the voltage is zero or negative.

The typical structure of the data frame, used for RS-232 applications, the first bit is the start bit, followed by the data bits, with the least significant bit first. The data bits may be in a packet of 5, 6, 7 or 8 bits. After the last data bit, there is an optional parity bit (even, odd or none) followed by a stop bit. Following the stop bit, there is a marking state of 1, 1½ or 2 bit periods, to indicate that the sequence of data bits is complete, before the next frame can be sent.

The capacitance in the connecting cable limits the maximum distance of transmission under RS-232 to a typical distance of 15 m (or 2500 pF). Excessive capacitance distorts the clean digital signal.

6.3.3   Interface mechanical characteristics

Although not specified by RS-232-C, the DB-25 connector (25-pin, D-type) has become so closely associated with RS-232 that it is the *de facto* standard. The DB-9 connector *de facto* standard (9-pin, D-type) is also commonly used. The pin allocations commonly used with the DB-9 and DB-25 connectors for the EIA-RS-232-C interface are shown in Appendix B. The pin allocation for the DB-9 connector is not quite the same as the DB-25; but is as follows:

- Data transmit   :   Pin 3
- Data receive   :   Pin 2
- Signal common   :   Pin 5

6.3.4   Functional description of the interchange circuits

The EIA circuit functions are defined, with reference to the DTE, as follows:

- **Pin 1 – protective ground (shield)**
  A connection is seldom made between the protective ground pins at each end. Their purpose is to prevent hazardous voltages, by ensuring that the DTE and DCE chassis are at the same potential at both ends. But, there is a danger that a path could be established for circulating earth currents. So, usually the cable shield is connected at one end only.

- **Pin 2 – transmitted data (TXD)**
  This line carries serial data from Pin 2 on the DTE to Pin 2 on the DCE. The line is held at MARK (or a negative voltage) during periods of line idle.

- **Pin 3 – received data (RXD)**
  This line carries serial data from Pin 3 on the DCE to Pin 3 on the DTE.

- **Pin 4 – request to send (RTS)**
  See clear to send.

- **Pin 5 – clear to send (CTS)**
  When a half-duplex modem is receiving, the DTE keeps RTS inhibited. When it becomes the DTEs turn to transmit, it advises the modem by asserting the RTS pin. When the modem asserts the CTS it informs the DTE that it is now safe to send data. The procedure is reversed when switching from transmit to receive.

- **Pin 6 – data set ready (DSR)**
  This is also called DCE ready. In the answer mode, the answer tone and the DSR are asserted two seconds after the telephone goes off hook.

- **Pin 7 – signal ground (common)**
  This is the common return line for the data transmit and receive signals. The connection, Pin 7 to Pin 7 between the two ends, is always made.

- **Pin 8 – data carrier detect (DCD)**
  This is also called the received line signal detector. Pin 8 is asserted by the modem when it receives a remote carrier and remains asserted for the duration of the link.

- **Pin 20 – DTE ready (or data terminal ready)**
  DTE ready enables, but does not cause, the modem to switch onto the line. In originate mode, DTE ready must be asserted in order to auto dial. In answer mode, DTE Ready must be asserted to auto answer.

- **Pin 22 – ring indicator**
  This pin is asserted during a ring on the line.

- **Pin 23 – data signal rate selector (DSRS)**
  When two data rates are possible, the higher is selected by asserting Pin 23.

6.3.5 The sequence of asynchronous operation of the RS-232 interface

Asynchronous operation is arguably the more common approach when using RS-232 and will be examined in this section using the more complex half duplex data interchange. It should be noted that the half-duplex description is given as it encompasses that of full-duplex operation.

Figure 6.13 gives a graphical description of the operation with the initiating user terminal (or DTE) and its associated modem (or DCE) on the left of the diagram and the remote computer and its modem on the right.

The following sequence of steps occurs:

- The initiating user manually dials the number of the remote computer.
- The receiving modem asserts the ring indicator line (RI-Pin 22) in a pulsed ON/OFF fashion as per the ringing tone. The remote computer already has its data terminal ready line (or DTR-Pin 20) asserted to indicate that it is ready to receive calls. (Alternatively, the remote computer may assert the DTR line after a few rings.) The remote computer then sets its request to send line (RTS-Pin 4) to ON.
- The receiving modem then answers the telephone and transmits a carrier signal to the initiating end. It also asserts the DCE ready (DSR-Pin 6) after a few seconds.
- The initiating modem then asserts the data carrier detect line (DCD-Pin 8). The initiating terminal asserts its DTR (if it is not already high). The modem then responds by asserting its data set ready line (DSR-Pin 6).
- The receiving modem then asserts its clear to send line (CTS-Pin 5) which permits the transfer of data from the remote computer to the initiating side.
- Data is then transferred from the receiving DTE on Pin 2 (transmitted data) to the receiving modem. The receiving remote computer can then transmit a short message to indicate to the originating terminal that it can proceed with the data transfer. The originating modem transmits the data to the originating terminal on Pin 3.
- The receiving terminal then sets its request to send line (RTS-Pin 4) to OFF. The receiving modem then sets its clear to send line (CTS-Pin 5) to OFF as well.
- The receiving modem then switches its carrier signal OFF.
- The originating terminal detects that the data carrier detect signal has been switched OFF on the originating modem and then switches its RTS line to the ON state. The originating modem then indicates that transmission can proceed by setting its CTS line to ON.
- Transmission of data then proceeds from the originating terminal to the remote computer.

- When the interchange is complete, both carriers are switched OFF (and in many cases the DTR is set to OFF). This means that the CTS, RTS and DCE ready (or DSR) lines are set to OFF.

Note that full-duplex operation requires that transmission and reception occur simultaneously. In this case, there is no RTS/CTS interaction at either end. The RTS line and CTS line are left ON with a carrier to the remote computer.

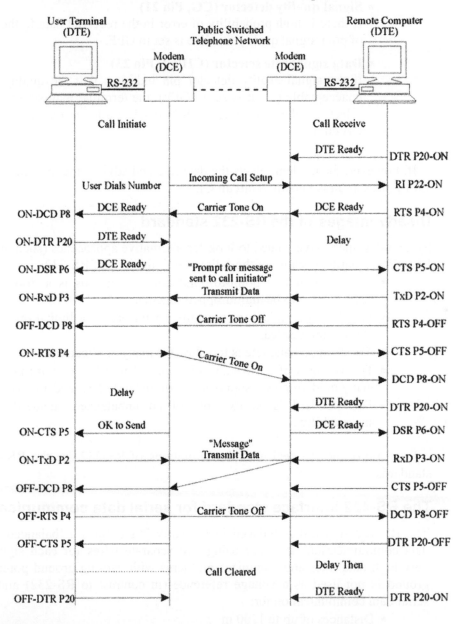

Figure 6.13
Typical operation of a half-duplex RS-232 data interchange

6.3.6   Synchronous communications

The major difference between asynchronous and synchronous communications with modems is the need for timing signals.

A synchronous modem outputs a square wave on Pin 15 of the RS-232 DB-25 connector. This Pin 15 is called the transmit clock pin or more formally the DCE transmitter signal element timing pin. This square wave is set to the frequency of the modem's bit rate. The attached personal computer (the DTE) then synchronizes its transmission of data from Pin 2 to the modem.

There are two interchange circuits that can be employed to change the operation of the attached communications device. These are:

- **Signal quality detector (CG, Pin 21)**
  If there is high probability of error in the received data to the modem because of poor signal quality this line is set to OFF.

- **Data signal rate selector (CH/CI, Pin 23)**
  If the signal quality detector pin indicates that the quality of the signal is unacceptable (i.e. it is set to OFF), the terminal may set the Pin 23 to ON to select a higher data rate; or OFF to select a lower data rate. This is called the CH circuit.

If, however, the modem selects the data rate and advises the terminal on Pin 23 (ON or OFF), the circuit is known as circuit CI.

6.3.7 Disadvantages of the RS-232 standard

System designers have tended to look for alternative approaches (such as the RS-422 and RS-485 standards) because of the following limitations of RS-232:

- The restriction of point-to-point communications is a drawback when many devices have to be multidropped together.
- The distance limitation (typically 15 meters) is a limitation when distances of 1000 m are needed.
- The 20 kbps baud rate is too slow for many applications.
- The voltages of −3 to −25 volts and +3 to +25 volts are not compatible with many modern power supplies (in computers) of +5 and +12 volt.
- The standard is an example of an unbalanced standard with high noise susceptibility.

Two approaches to deal with the limitations of RS-232 are the RS-422 and RS-485 standards.

6.3.8 The RS-422 interface standard for serial data communications

The RS-422 standard introduced in the early 70s defines a 'balanced' (or differential) data communications interface using two separate wires for each signal. This permits very high data rates and minimizes problems with varying ground potential because the ground is not used as a voltage reference (in contrast to RS-232) and allows reliable serial data communication for:

- Distances of up to 1200 m
- Data rates of up to 10 Mbps
- Only one line driver is permitted on a line
- Up to 10 line receivers can be driven by one line driver

The line voltages range between –2 V to –6 V for Logic 1 and +2 V to +6 V for Logic 0 (using terminals A and B as reference points). The line driver for the RS-422 interface produces a ±5 V differential voltage on two wires.

The two signaling states of the line are defined as follows:

- When the 'A' terminal of the driver is negative with respect to the 'B' terminal, the line is in a binary 1 (Mark or Off) state.
- When the 'A' terminal of the driver is positive with respect to the 'B' terminal, the line is in a binary 0 (Space or On) state.

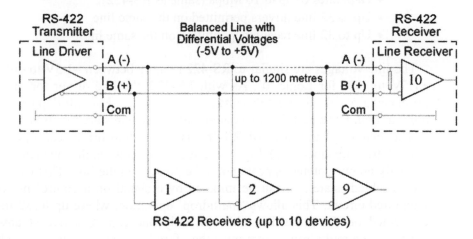

Figure 6.14
The RS-422 balanced line driver connections

As the differential receiver is only sensitive to the difference between two signals on its inputs, common noise signals picked up in both wires will have little effect on the operation of the receiver. Differential receivers are therefore said to have good common mode rejection (CMR) characteristics. The voltage drop along the common wire of an unbalanced system due to other signals is also eliminated.

The differential voltage signal is the major feature of the RS-422 standard, which allows an increase in speed and provides higher noise immunity. Each signal is transferred on one pair of wires and is the voltage difference between them. A common ground wire is preferred to aid noise rejection. Consequently, for a full-duplex system, five wires are required (with 3 wires for half-duplex systems).

The balanced line driver can also have an input signal called an 'enable' signal. The purpose of this signal is to connect the driver to its output terminals, A & B. If the enable signal is off, one can consider the driver as disconnected from the transmission line or in a high impedance state. This 'tri state' approach (logic '0', logic '1' and high impedance) is effectively the RS-485 standard (discussed next). The differential lines of the RS-422 are normally terminated with a resistor equal to the characteristic impedance (Z_o) of the line. This will prevent signal distortion due to reflections from the end of line. A typical value of Z_o would be in the order of 120 ohms.

The RS-422 standard does not specify the mechanical connections or assign pin numbers and leaves this aspect optional. The RS-530 for the DB-25 connector as shown in Appendix B is sometimes used.

6.3.9 The RS-485 interface standard for serial data communications

The RS-485 standard is the most versatile of the four EIA interface standards discussed in this chapter. It is an expansion of RS-422 and allows the same distance and data speed but increases the number of transmitters and receivers permitted on the line. RS-485 permits multidrop network connection on two wires and provides for reliable serial data communication for:

- Distances of up to 1200 m (same as RS-422)
- Data rates of up to 10 Mbps (same as RS-422)
- Up to 32 line drivers permitted on the same line
- Up to 32 line receivers permitted on the same line

The line voltages are similar to RS-422 ranging between −1.5V to −6V for logic '1' and +1.5V to +6V for Logic '0'. As with RS-422, the line driver for the RS-485 interface produces a 5 volt differential voltage on two wires. For full-duplex systems, five wires are required. For a half-duplex system, only three wires are required.

The major enhancement of RS-485 is that a line driver can operate in three states (called tri-state operation) logic '0', logic '1' and 'high-impedance', where it draws virtually no current and appears not to be present on the line. This latter state is known as the 'disabled' state and can be initiated by a signal on a control pin on the line driver integrated circuit. This allows 'multidrop' operation, where up to 32 transmitters can be connected on the same line, although only one can be active at any one time. Each terminal in a multidrop system must therefore be allocated a unique address to avoid any conflict with other devices on the system. RS-485 includes current limiting in cases where contention occurs.

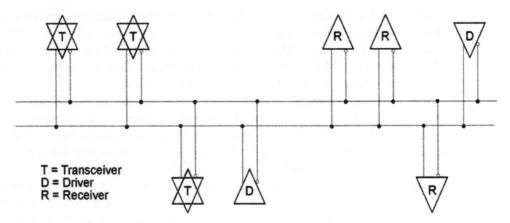

Figure 6.15
The RS-485 multipoint interface standard

The RS-485 interface standard is useful where distance and connection of multiple devices on the same pair of lines is desirable. Special care must be taken with the software to coordinate which devices on the network can become active. In most cases, a master terminal, such as a PLC or computer, controls which transmitter/receiver will be active at any one time.

On long lines, the leading and trailing edges of data pulses will be much sharper if terminating resistors approximately equal to the characteristic impedance (Z_o) of the line are fitted at the extreme ends.

6.4 Flow control

These techniques are widely used in modem communications to ensure that there will be no overflow of data by the device receiving a stream of characters, which it is temporarily unable to process, or store. The receiving device needs a facility (called flow control) to signal to the transmitter to temporarily cease sending characters down the line.

There are three mechanisms of flow control described below. The first two techniques are software based and the last, RTS/CTS, is hardware-based handshaking.

- **XON/XOFF signaling**

  When the modem decides that it has too much data arriving, it sends an XOFF character to the connected terminal to tell it to stop transmitting characters. The need to advice the terminal to stop transmitting characters is typically when the modem memory buffer fills up to 66% full. This delay in transmission of characters by the terminal allows the modem to process the data in its memory buffer. Once the data has been processed (and the memory buffer has emptied to typically 33% full), the modem sends an XON character to the terminal and transmission of data to the modem then resumes. XON and XOFF are two defined ASCII characters DC1 and DC3 respectively. This system works well unless there are flow control characters (XON/XOFF) in the normal data stream. This can cause problems and these characters should be removed from the standard stream of information transmitted and reserved for control purposes.

- **ENQ/ACK**

  The terminal sends an ENQ control character to the modem when it wants to transmit a finite block of data. When the modem is ready to receive characters it transmits an ACK which then allows the terminal to commence transmission of this block of data. The process is repeated for subsequent blocks of data.

- **RTS/CTS signaling**

  This technique of hardware flow control is a simplified version of the full hardware handshaking sequence discussed earlier. When the terminal wants to transmit data to the modem, it asserts the request to send (RTS) line and waits for the modem to assert the clear to send (CTS) line before transmitting. When the modem is unable to process any further characters it switches off (or inhibits) the CTS control line. The terminal device then stops transmitting characters until the CTS line is asserted again.

6.5 Modulation techniques

In essence, the modulation process modifies the characteristics of a carrier signal. The carrier signal can be represented as a sine wave:

$$f(t) = A \sin (2 \pi \, ft + \cos \phi)$$

where:

f(t)	=	*instantaneous value of voltage at time t*
A	=	*maximum amplitude*
f	=	*frequency*
phi	=	*phase angle*

The various modulation techniques will now be examined.

6.5.1   Amplitude modulation (or amplitude shift keying)

The amplitude of the carrier signal is varied in correspondence with the binary stream of data coming in.

ASK or amplitude shift keying is sometimes still used for low data rates; however, it does have problems with distinguishing the signal from noise in the communications channel, as noise is amplitude based phenomena.

6.5.2   Frequency modulation (or frequency shift keying – FSK)

This approach allocates different frequencies to the logic 1 and logic 0 of the binary data messages. This is primarily used by modems operating at data rates up to 300 bits per second (bps) in full-duplex mode and 1200 bps in half-duplex mode.

The Bell 103/113 and the compatible CCITT V.21 standards are indicated in Table 6.2

Specification	Originate (Mark)	Originate (Space)	Answer (Mark)	Answer (Space)
CCITT V.21	980 Hz	1180 Hz	1650 Hz	1850 Hz
Bell 103	1270 Hz	1070 Hz	2225 Hz	2025 Hz

Table 6.2
CCITT V.21 and Bell system 103/113 modems frequency allocation

The Bell 103/113 modems had to be set up in either 'originate' or 'answer mode'. Typically, terminals were connected to originate modems and main frame computers were connected to answer type modems. It is thus easy to communicate when originate modems are connected to answer mode modems. Similar modems cannot communicate with each other as they expect different frequencies (e.g. two originate modems connected together).

Because of the two different bands of frequencies in which the sets of signals operate full duplex operation is possible with these modems. Note that they fit into the allowable bandwidth of the communications channel.

6.5.3   Phase modulation (or phase shift keying (PSK))

This is the process of varying the carrier signal by phase. There are various forms of phase modulation. In quadrature (four phases) phase shift keying (QPSK) four phase angles are used for encoding: 0°, 90°, 180° and 270° as indicated in Figure 6.16.

There are four phase angles possible at any possible time; thus allowing the basic unit of data to be a 2-bit pair (or dibit). The weakness with this approach is that a reference signal is required as per Figure 6.16.

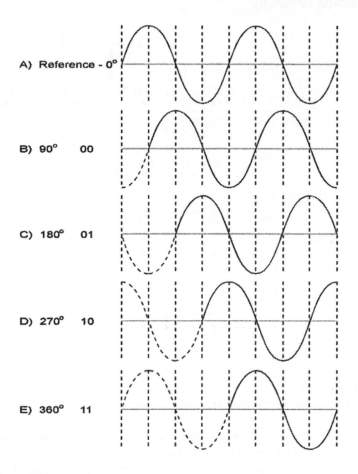

Figure 6.16
Quadrature phase shift keying

The preferred option is thus to use differential phase shift keying where the phase angle for each cycle is calculated relative to the previous cycles as shown in Figure 6.17.

A modulation rate of 600 baud results in a data rate of 1200 bits per second using two bits for each phase shift.

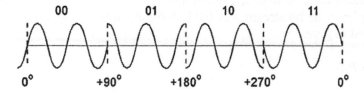

Figure 6.17
Differential phase shift keying

6.5.4 Quadrature amplitude modulation (or QAM)

Two parameters of a sinusoidal signal (amplitude and phase) can be combined to give QAM. This allows for 4 bits to be used to encode every amplitude and phase change. Hence a signal at 2400 baud would provide a data rate of 9600 bps. The first implementation of QAM provided for 12 values of phase angle and 4 values of amplitude.

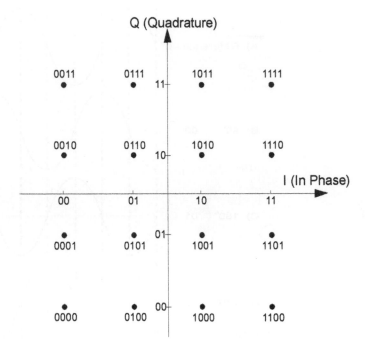

Figure 6.18
CCITT V.22bis quadrature amplitude modulation

QAM also uses two carrier signals. The encoder operates on 4 bits for the serial data stream and causes both an in-phase (IP) cosine carrier and a sine wave that serves as the quadrature component (QC) of the signal to be modulated. The transmitted signal is then changed in amplitude and phase resulting in the constellation pattern above.

6.5.5 Trellis coding

QAM modems are susceptible to noise; hence, a new technique called trellis coding was introduced. These allow for 9600 to 56 k bps transmission over the normal telecom lines. In order to minimize the errors that occur when noise is evident on the line, an encoder adds a redundant code bit to each symbol interval.

For example, if the bit stream was 1011 ($b_4b_3b_2b_1$) it will have an additional four check bits calculated to give:

$P_4b_4P_3b_3P_2b_2P_1b_1$
where:
$P_1 = b_1\ XOR\ b_0 = 1\ XOR\ 0 = 1$
$P_2 = b_2\ XOR\ b_1 = 1\ XOR\ 1 = 0$
$P_3 = b_3\ XOR\ b_2 = 0\ XOR\ 1 = 1$
$P_4 = b_4\ XOR\ b_3 = 1\ XOR\ 0 = 1$

This gives a sequence of:
11100111

Only certain sequences are valid. If there is noise on the line, which causes the sequence to be different from an accepted sequence, the receiver will then select the valid signal point closest to the observed signal without needing a retransmission of the affected data.

A typical comparison in performance would be that of a conventional QAM modem that might require 1 of every 10 data blocks to be retransmitted, could be replaced by a modem using trellis coding with only one in every 10 000 data blocks to be in error.

6.5.6   DFM (direct frequency modulation)

This is mentioned as a separate form of modulation for completeness. It is one method of modulating digital information with an analog modulator. However it can be considered to be a form of FSK. It is widely referred to in radio communications. The correct technical name is 'Gaussian minimum shift keying' (GMSK)

It is possible to directly modulate the data directly onto the radio frequency carrier. This cannot be done within the normal telephone lines because of bandwidth limitations (3 kHz as opposed to radio channel bandwidths of 12.5 kHz). Simple filtering of the square wave (binary) data signal ensures that the channel spectrum is indeed limited to 12.5 kHz.

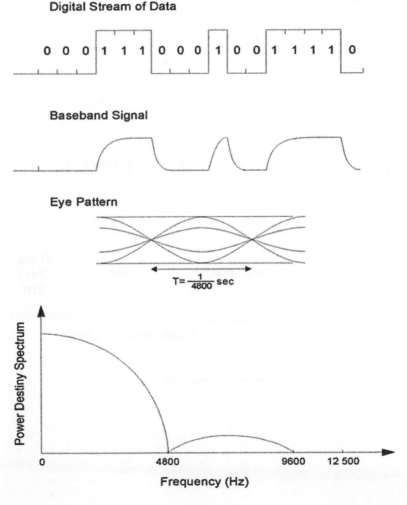

Figure 6.19
Direct frequency modulation

This method of modulation is used in some radio systems that claim to be 'digital radio'. In reality, it is a hybrid of digital and analog systems and not a true digital transmission technique.

6.6 Error detection/correction and data compression

The most popular forms of error detection were initially cyclic redundancy checks (especially the CRC-16). These are discussed in detail in Chapter 2 under *Error detection and correction*. Unfortunately different vendors implemented minor variations on the CRC approach, which resulted in incompatibilities between the different vendor's products. The advent of the Microcom networking protocol (MNP), licensed by Microcom to numerous other manufacturers resulted in a *de facto* standard developing.

6.6.1 MNP protocol classes

There are nine MNP protocol classes defined in Table 6.3 below. Modems that connect together agree with each other on the highest supported class of MNP to transfer data at. An initial frame called the link request is used to establish the standards to be followed in transferring the data.

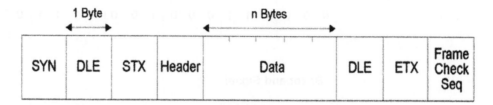

Asynchronous Data Frame

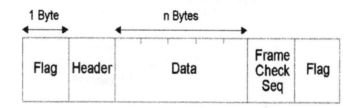

Synchronous Data Frame

Figure 6.20
Asynchronous and synchronous MNP frame formats

MNP Class	Async/ Synchronous	Half or Full Duplex	Efficiency	Description
1	Asynchronous	Half	70%	Byte oriented protocol
2	Asynchronous	Full	84%	Byte oriented protocol
3	Synchronous	Full	108%	Bit oriented protocol. Communications between (PC) terminal and modem is still asynchronous
4	Synchronous	Full	120%	Adaptive Packet Assembly (large data packets used if possible). Data phase optimization (elimination of protocol administrative overheads)
5	Synchronous	Full	200%	Data Compression rates of 1.3 to 2.0
6	Synchronous	Full	–	9600 bps V.29 modulation universal link negotiation allows modems to locate the highest operating speed and use statistical multiplexing
7	Synchronous	Full	–	Huffman encoding (enhanced data compression) reduces data by 42%
8	Synchronous	Full	–	CCITT V.29 Fast Train Modem technology added to class 7
9	Synchronous	Half Duplex emulates Full Duplex	–	CCITT V.32 modulation + Class 7 Enhanced data compression. Selective retransmission in which error packets are retransmitted.

Table 6.3
MNP protocol classes

6.6.2 Link access protocol modem (LAP-M)

This is recognized as the primary method for error detection and correction under the CCITT V.42 recommendation (MNP error detection and correction is the secondary mechanism).

The various data compression techniques are discussed in greater detail below.

6.6.3 Data compression techniques

The main reason for data compression is to achieve higher effective speeds in the transmission of the data (and a reduction in transmission times).

Two of the most popular data compression methods are adaptive computer technologies compressor technology and Microcom's MNP class 5 and class 7 compression procedures. In 1990, the CCITT promulgated the V.42 bis standard, which defines a new data compression method known as Lempel-Ziv.

The three compression standards that will be discussed here are:

6.6.3.1 MNP class 5 compression

This is a two-stage process:

- **Run length encoding**

  The first three bytes indicate the beginning of a run length encoded sequence. The next byte is the repetition count of bits (with a maximum number of 250 bits). For runs of similar bits this can reduce the total size of the data bytes dramatically.

  An example of run length encoding is for the use of fax machines (modified modems).

  Essentially the number of successive bits, which are the same are counted and then coded into an eight-bit symbol, for example. The eight-bit symbol is then transmitted.

  Data compression is discussed with reference to the well-known fax machine. For example in Group 3 machines, a regular 11 inch sheet of paper can be vertically digitized into 100 lines per inch to produce 1100 lines and horizontally each line is further digitized into 1700 bits/line.

  Total size of the file = 1700 bits/line × 1100 lines = 1.87 Mbits

  Assuming this file is sent on a 2400 baud modem, the transmission time for one page of text would be 779 seconds, as calculated below. However, in practice, the transmission time of a page is about 30 to 60 secs. Data compression is used to achieve these results.

  $$Transmission\ time = \frac{(1.87\ \text{Mbits})}{(2400\ \text{bits/second})}$$

  $$= 779\ \text{seconds}$$

  The microprocessor on the facsimile machine can process the data bits before sending them and uses a compression algorithm for compressing the data into fewer bits.

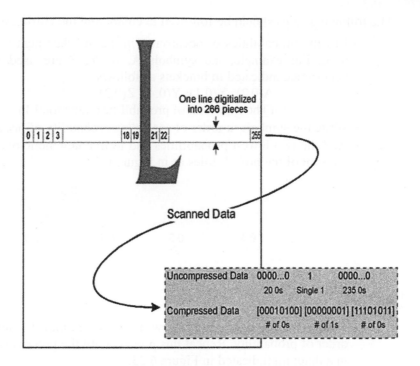

One line digitialized into 266 pieces

| 0 | 1 | 2 | 3 | | 18 | 19 | | 21 | 22 | | 255 |

Scanned Data

Uncompressed Data	0000...0	1	0000...0
	20 0s	Single 1	235 0s
Compressed Data	[00010100]	[00000001]	[11101011]
	# of 0s	# of 1s	# of 0s

Figure 6.21
Data compression techniques applied to a scanned line

- **Adaptive frequency coding**
  A compression token is substituted for the actual byte transferred. Shorter tokens are substituted for more frequently occurring data bytes. A compression token consists of two parts:
  - A fixed length header (3-bit long indicating length of body)
  - A variable length body

At compression initialization, a table is set up for each byte from 0 to 255. To encode a data byte, the token to which it is mapped is substituted for the actual data byte in the data stream. The frequency of occurrence of the current data byte is incremented by one. If the frequency of occurrence of the current data byte is greater than the frequency of the next most frequently occurring data byte, the two tokens are swapped. This comparison process is then repeated for the next most frequently occurring data byte and the tokens swapped again.

6.6.3.2   MNP class 7 enhanced data compression

This combines run length encoding with an adaptive encoding table. This table is used to predict the probability of a character based on the value of the previous character. Up to 256 coding tables are kept for each 8-bit pattern ($2^8 = 256$). All characters are organized according to the rules of Huffman coding.

The principle of Huffman encoding relies on some characters occurring more frequently than others. The Huffman code is computed by the computer determining the frequency of occurrence of each symbol (in the set of symbols used for communications).

The following steps should be followed in computing the Huffman codes:

- List the probabilities of occurrence in the message next to each of the symbols used. For example, the symbols A, X, Y, Z are used with probabilities of occurrence indicated in brackets as follows:
  A(0.2); X(0.1); Y(0.4); Z(0.3)
  (The sum total of probabilities must total 1)
- Write the symbols in order of ascending probability of occurrence.
- Add the two lowest probabilities and form a new node over the two nodes with the sum of the probabilities as in Figure 6.22.

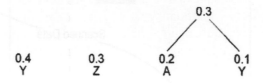

Figure 6.22
First node

- Repeat the process with the new node created and the next node to the left in order of probability. Repeat this process until the process is completed, resulting in a diagram indicated in Figure 6.23.

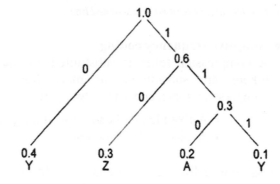

Figure 6.23
Second and third nodes

- Assign a 1 to the branches leaning in one direction as indicated above and 0 to the remaining branches.
- Compute the Huffman code for each symbol by tracing the path from the apex of the pyramid to each base.

Hence:

$Y = 0$
$Z = 10$
$A = 110$
$X = 111$

In order to compute the compression ratio (as compared to the standard 7-bit ASCII code) assume there are 1000 symbols (i.e. Y, Z, A, and X) transmitted.

Total bits using Huffman encoding =
(Probability of occurrence of symbol of 0.4 * 1000 symbols) * 1 bit/symbol Y +
(Probability of occurrence of symbol of 0.3 * 1000 symbols) * 2 bits/symbol Z +
(Probability of occurrence of symbol of 0.2 * 1000 symbols) * 3 bits/symbol A +
(Probability of occurrence of symbol of 0.1 * 1000 symbols) * 3 bits/symbol X
= 400 + 600 + 600 + 300
= 1900 bits

If the ASCII code had been used this would have resulted in –
1000 symbols * 7 bits/symbol = 7000 bits
Hence the compression ratio = 7000/1900 = 3.68.

- Once the Huffman code has been computed, the software converts each symbol into its equivalent code and includes the table used for translating the code back to the original symbols in the original transmission. The receiver software will then decompress the stream of bits into the original stream of symbols. Run length encoding is used if there are four or more identical characters in a specified sequence of characters. The first three characters are encoded (as for the rules of Huffman encoding) and the number of remaining (identical) characters encoded in a 4-bit nibble. Decoding the data stream is done quite simply because the receiving modem keeps the same compression table as the transmitting modem.

6.6.3.3.   V.42bis

This relies on the construction of a dictionary, which is continually modified as the data is transferred between the two modems. The dictionary consists of a set of trees in which each root corresponds to a character in the alphabet. When communications is established each tree comprises a root node with a unique code word assigned to each node. The sequence of characters received by the modem (from its attached terminal) is compared and then matched against the dictionary.

The maximum string length can vary from 6 to 250 characters and is defined by the two connecting modems. The number of code words has a minimum of 512 (but any value above this default minimum value can be agreed to between the two connecting modems).

V.42 bis data compression, in substituting a code word for a string, is between 20 and 30% more efficient than MNP class 5 compression. V.42 bis is effective for large file transfers, but not for short strings of data.

6.7   Data rate versus baud rate

Baud rate is the physical signaling rate on the communications link. In a simple application, there is only one signal change to represent each bit. When the signal change represents two bits, this is called dibit encoding. Similarly using 1 baud rate to represent 3 bits is known as tribit encoding. Phase modulation is the preferred technique to achieve multiple bit encoding. For example, if there are four possible phase angles a particular signal can take, dibit encoding can be used where 2 bits are transmitted for every signal change.

Frequency shift keying on the other hand has normally only two frequencies possible. Hence, each signal change has one bit associated with it (1 baud = 1 bps).

6.8   Modem standards

A table, which summarizes the various CCITT modem standards, is given below.

Modem Type	Data Rate	Async / Sync	Mode	Modulation	Switched / Leased
V.21	300	Async	Half / Full	FSK	Switched
V.22	600	Async	Half / Full	DPSK	Switched / Leased
	1200	Async / Sync	Half / Full	DPSK	Switched / Leased
V.22 bis	2400	Async	Half / Full	QAM	Switched
V.23	600	Async / Sync	Half / Full	FSK	Switched
	1200	Async / Sync	Half / Full	FSK	*Switched*
V.26	2400	Sync	Half / Full	DPSK	Leased
	1200	Sync	Half	DPSK	Switched
V.26 bis	2400	Sync	Half	DPSK	Switched
V.26 ter	2400	Sync	Half / Full	DPSK	Switched
V.27	4800	Sync	Full	DPSK	Leased
V.27 bis	4800	Sync	Full	DPSK	Leased
	2400	Sync	Full	DPSK	Leased
V.27 ter	4800	Sync	Half	DPSK	Switched
	2400	Sync	Half	DPSK	Switched
V.29	9600	Sync	Half / Full	QAM	Leased
V.32	9600	Async	Half / Full	TCM / QAM	Switched
V.33	14400	Sync	Half / Full	TCM	Leased

Table 6.4
Modem standards

The following table summarizes the various Bell modem standards.

Modem Type Bell System	Data Rate	Transmission Technique	Modulation Technique	Transmission Mode	Line Use
103 A, E	300	Asynchronous	FSK	Half, Full	Switched
103 F	300	Asynchronous	FSK	Half, Full	Leased
201 B	2400	Synchronous	PSK	Half, Full	Leased
201 C	2400	Synchronous	PSK	Half, Full	Switched
202 C	1200	Asynchronous	FSK	Half	Switched
202 S	1200	Asynchronous	FSK	Half	Switched
202 D/R	1800	Asynchronous	FSK	Half, Full	Leased
202 T	1800	Asynchronous	FSK	Half, Full	Leased
208 A	4800	Synchronous	PSK	Half, Full	Leased
208 B	4800	Synchronous	PSK	Half	Switched
209 A	9600	Synchronous	QAM	Full	Leased
212	0-300	Asynchronous	FSK	Half, Full	Switched
	1200	Asynchronous/ Synchronous	PSK	Half, Full	Switched

Table 6.5
Bell modem standards

6.9   Radio modems

Radio modems are suitable for replacing wire lines to remote sites or as a backup to wire or fiber-optic circuits.

Figure 6.24
Radio modem configuration

Modern radio modems operate in the 400 and 900 MHz band. Propagation in the 400 and 900 MHz band requires a free line of sight between transmitting and receiving antennae for reliable communications. They can be operated in a network, but require a network management software system (protocols) to manage the network access and error detection. Very often, a master station (with hot changeover) communicates with multiple radio field stations. The protocol for these applications can use a simple poll/response technique.

For the more sophisticated peer-to-peer network communications applications, a protocol based on CSMA/CD (carrier sensing multiple access with collision detection) would be necessary. A variation on the standard approaches is to use one of the radio modems as a network watchdog to periodically poll all the radio modems on the network and to check their integrity. The radio modem can also be used as a relay station to communicate to other systems, which are out of the range of the master station. These issues are discussed in Chapter 2.8 under communication philosophies.

These devices are designed to ensure that computers and PLCs, for example, can communicate transparently over a radio link without any specific modifications required.

The interface to the radio modem is typically RS-232 but RS-422 and RS-485 (and fiber optics) is provided as options. Typical speeds of operation are up to 9600 baud. A buffer is required in the modem (typically a minimum size of 32 kbps). The various hardware and software flow control techniques (as discussed in Section 6.4) are normally provided in the radio modem firmware for ensuring that there is no loss of data between the radio modem and the connecting terminal.

Typical modulation techniques are two level direct FM (1200 to 4800 bps) to three level direct FM (9600 bps).

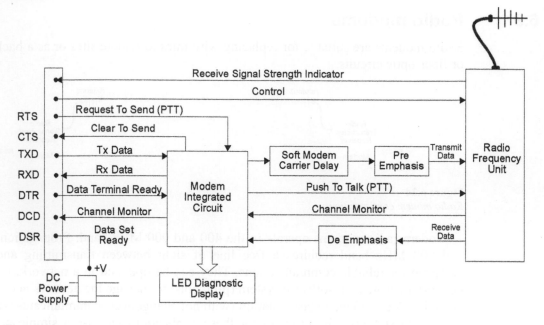

Figure 6.25
Typical structure of a radio modem

A typical schematic of a radio modem is given in Figure 6.25. The following terms are used in the diagrams relating to radio modems:

- **PTT**
  Push to talk signal

- **RSSI**
  Receive signal strength indicator indicates the received signal strength with a proportionally varying DC voltage

- **Noise squelch**
  This attempts to minimize the reception of any noise signal at the discriminator output.

- **RSSI squelch**
  This opens the 'receive audio' path when the signal strength of the RF carrier is of sufficiently high level.

- **Channel monitor**
  This indicates if the squelch is open.

- **Soft carrier delay**
  This allows for the RF transmission to be extended slightly after the actual end of the data message. This avoids the end of transmission bursts that occur when the carrier stops and the squelch (almost) simultaneously disconnects the audio path.

The RTS, CTS, DCD, clock, transmit data, receive data all relate to the RS-232 section discussed in Section 6.3.1.

The radio modem has a basic timing diagram indicated in Figure 6.26 for communications between a terminal and the radio modem.

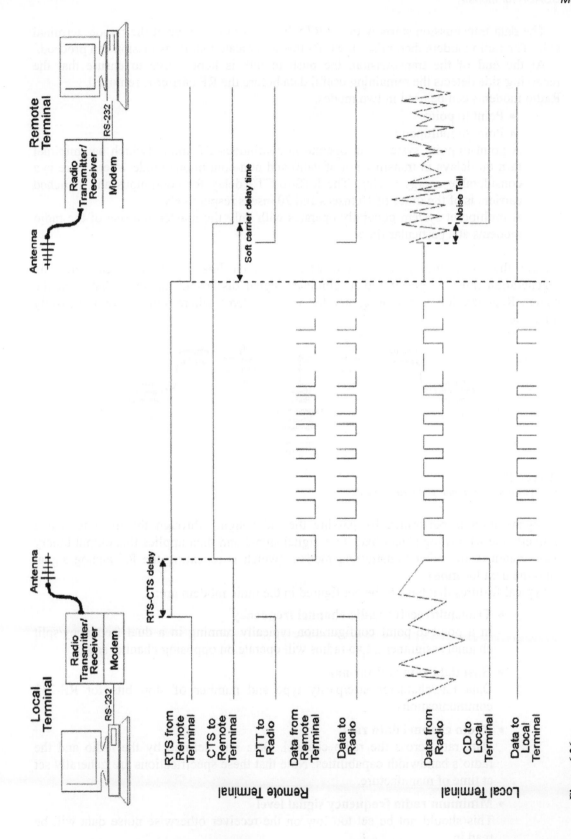

Figure 6.26
Radio modem timing diagram

The data transmission starts with the RTS line becoming active at the remote terminal side. The radio modem then raises the CTS line to indicate that transmission can proceed.

At the end of the transmission, the push to talk is kept active to ensure that the receiving side detects the remaining useful data before the RF carrier is removed.

Radio modems can be used in two modes:

- Point to point
- Point to multipoint

A point-to-point system can operate in continuous RF mode (which has minimal turn on delays in transmission of data) and non-continuous mode where there is a considerable energy saving. The RTS-to-CTS delay for continuous and switched carriers is of the order of 10 msecs and 20 msecs respectively.

A multipoint system generally operates with only the master and one of the radio modems at a particular time.

Note that in a multipoint system when the data link includes a repeater, data regeneration must be performed to eliminate signal distortion and jitter that is in the signal. Regeneration is not necessary for voice systems where some error is possibly tolerable.

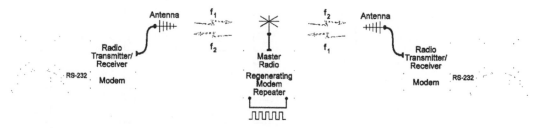

Figure 6.27
Regeneration of a signal with a radio modem

Regeneration is performed by passing the radio signal through the modem, which converts the RF analog signal back to a digital signal and then applies this output binary data stream to the other transmitting modem (which is repeating the RF analog signal onto the next location).

Typical features that have to be configured in the radio modem are:

- **Transmit/receive radio channel frequency**
  In a point-to-point configuration typically running in a dual frequency/split channel assignment, two radios will operate on opposing channel sets.

- **Host data rate and format**
  Data rate/character size/parity type and number of stop bits for RS-232 communications.

- **Radio channel data rate**
  Data rate across the radio channel. Data rate defined by the radio and the radio's bandwidth capabilities. Note that these specifications are generally set at time of manufacture.

- **Minimum radio frequency signal level**
  This should not be set too low on the receiver otherwise noise data will be read in.

- **Supervisory data channel rate**
  This is used for flow control – hence should not be set too low otherwise the buffer on the receiver will overflow. Typically one (1) flow control bit to thirty-two bits of serial data is a standard.

- **Transmitter key up delay**
  This is the time for the transmitter to energize and stabilize before useful data is sent over the radio link. This should be kept as low as possible to minimize on overheads.

Several countries around the world have allocated a section of bandwidth for use with 'Spread Spectrum' radio modems. In America and Australia this is in the 900 MHz area.

Briefly, a very wideband channel is allocated to the modem. In this case approximately 3.6 MHz wide. The transmitter uses a pseudo random code to place individual bits (or groups of bits) broadly across the bandwidth and the receiver uses the same random code to receive them. Because they are random, a number of transceivers can operate on the same channel and a collision of bits will be received as noise by a close by receiver.

The advantage of this approach is very high data security and data speeds up to 19.2 kbps. The disadvantage is the very inefficient use of the radio spectrum.

6.10   Troubleshooting the system

The two aspects of troubleshooting a communications system (as indicated in Figure 6.28) will be examined here.

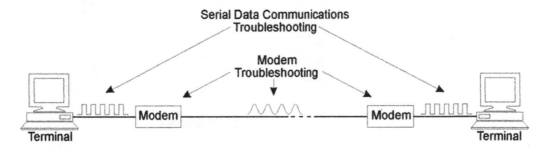

Figure 6.28
Troubleshooting a system

6.10.1   Troubleshooting the serial link

When troubleshooting a serial data communications system, a logical approach needs to be followed to avoid wasting too much time.

A few suggestions are:

- Check the basic parameters such as are the baud rate, data format and data format correct for both communicating devices.
- Identify which is the DTE and which is the DCE device and ensure that the transmit pin is connected to the receiver pin at the other end.
- Clarify what is happening with the hardware handshaking.
- Examine the protocol being used if the first few suggestions above do not help.

The following three useful devices can be used to assist in analyzing the problem:

- Digital multimeter

- Breakout box
- Protocol analyzer (PC based is a useful solution)

The breakout box and the protocol analyzer are briefly discussed in the following paragraphs.

6.10.2   The breakout box

The breakout box is an inexpensive tool that provides most of the information necessary to identify and fix problems on data communications circuits, such as the serial RS-232, RS-422, RS-423, RS-485, etc interfaces and also on parallel interfaces.

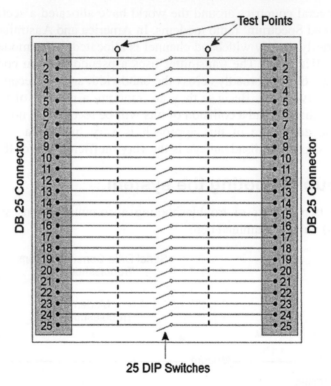

Figure 6.29
Breakout box showing test points

A breakout box is connected into the data cable, to bring out all 25 (or 9, 37, 50, etc) conductors in the cable to accessible test points. Breakout boxes usually have a male and a female socket and by using 2 standard serial cables the box can be connected in series with communication link. The 25 test points can be monitored by LEDs, a simple digital multimeter, an oscilloscope, or a protocol analyzer. In addition, a switch in each line can be opened or closed while trying to identify where the problem is.

6.10.3   Protocol analyzer

A protocol analyzer is used to display the actual bits on the data line, as well as the ASCII control codes, such as XON, XOFF, LF, CR, etc. The protocol analyzer can be used to monitor the data bits as they are sent down the line and compared with what should be on the line. This helps to confirm that the transmitting terminal is sending the correct data and that the receiving device is receiving it. The protocol analyzer is useful in identifying incorrect setting of baud rate, parity, stop bit, noise or incorrect wiring and connection. It also makes it possible to analyze the format of the message and look for protocol errors.

When the problem has been shown not to be due to the connections, baud rate, bits or parity, then the content of the message will have to be analyzed for errors or inconsistencies. Protocol analyzers can quickly identify these problems.

6.10.4 Troubleshooting the modem

There are various tests open to troubleshooting a modem to identify any problems with the operation of it.

The first test is the self-test one where the modem connects its transmitter to its receiver. The connection with the communications line is broken and a specified sequence of bits is transmitted to the receiving parts of the modem where this is then compared with a defined pattern.

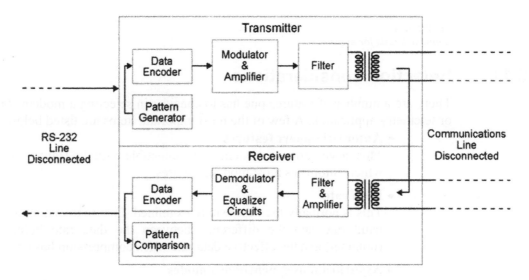

Figure 6.30
Modem internal self-test

An error will be indicated on the modem front panel if the transmitted sequence does not match the expected pattern.

The second set of tests is the loop back tests. There are four forms of loop back tests:

- Local digital loop (to test the terminal or computer and connecting RS-232 line)
- Local analog loop (to test the modem's modulator and demodulator circuitry)
- Remote analog loop (to test the connecting cable and local modem)
- Remote digital loop (to test the local and remote modem and connecting cable)
- The figure below gives a diagrammatic example of each approach.

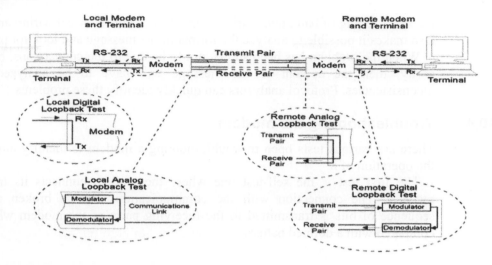

Figure 6.31
Loop back tests for modems

6.11   Selection considerations

There are a number of features one has to consider in selecting a modem for an industrial or telemetry application. A few of the most important items are listed below.

- **Automatic smart features**
  Most asynchronous modems are compatible with the Hayes AT command set, which automates most modem features.

- **Data rate**
  This is normally the first feature examined. Ensure to distinguish this from the baud rate and the difference between the data rate before compression (nominal) and the effective data rate when compression has been performed.

- **Asynchronous/synchronous modes**
  The ability to switch between both modes allows more flexibility for future applications. This is sometimes provided in a dip switch form.

- **Transmission modes**
  The most efficient method of data transfer is full-duplex. This is a preferable method of operation to half-duplex where the line turnaround time introduces a considerable amount of inefficiency into the data transfer.

- **Modulation techniques**
  The two most popular modulation techniques are V.22 bis which supports 1200 and 2400 bps transmission and V.32 bis the other almost universal transmission capability (which has V.22 bis as a subset).

- **Data compression techniques**
  There are four main compression standards used today (mainly for telecom switched lines). It is important that the modem has compatibility with these:

  - MNP class 5 (the most popular)
  - MNP class 7 (compatible with MNP 5)
  - ACT
  - CCITT V.42bis (arguably becoming the most popular)

- **Error correction/detection**
  The most popular error detection and correction mechanism is MNP-4. The CCITT have incorporated this standard into the V.42 standard, which allows for MNP-4 and link access procedure for modems (LAP-M).

- **Flow control**
  This is useful in controlling the flow of data from an attached terminal so that it does not overload the modem. Ensure that the existing terminals and hardware support the necessary flow control protocols such as ENQ/ACK, RTS/CTS or XON/XOFF.

- **Optimal blocking of data (or protocol spoofing)**
  Two modems would negotiate with each other for the specific file transfer protocol that should be used before transfer of data occurs. This avoids unnecessary acknowledgments from the connected terminal device to the modem. Hence if two modems can transfer 500 character blocks between them, but the terminal to modem only supports 100 character blocks, the modem would accumulate 5 sets of 100 character blocks and transfer this in one hit to the receiving modem. The receiving modem would then transfer 5 sets of 100 character blocks to the receiving terminal, which would then acknowledge each 100-character block in turn.

- **Rack mounted/internal/stand-alone modems**
  Selection should be made on the basis of the application. A lot of industrial systems use rack mounted modems for saving on space and ease in providing the appropriate power supplies.

- **Power supply**
  Typically most manufacturers of the latest modems have a separate power supply or derive power from the telephone lines.

- **Self testing features**
  Ensure that the modem can perform a self-test and also the standard local and remote loop back tests.

7

Central site computer facilities

7.1 Introduction

The central site computer facilities have to be designed and installed to ensure the satisfactory operation of the hardware and software and to ensure that the operators and other users can use the system effectively and safely.

This chapter discusses the requirements for the central site computer facilities with reference to the following:

- Recommended installation practice
- Ergonomic requirements
- Design of the computer displays
- Alarming and reporting philosophies

7.2 Recommended installation practice

There are a number of requirements, which have to be carefully adhered to in installing the computer system in a building. These are reviewed in the following paragraphs.

7.2.1 Environmental considerations

The environment in which the system is installed must be appropriate to the computer system and the associated electronics systems. Typical environmental conditions that are considered suitable for the standard and the industrial environment are listed in Table 7.1. Obviously, the environment in a control room should not have these extremes; but the equipment should be rated for these ranges. Typical control room environmental ranges are discussed under ergonomic requirements. Industrial computer systems may be mounted in a less stringent environment than for the standard air-conditioned control room.

Environmental Condition	Recommended Range	
	Industrial	**Standard**
Operating Temperature	0°C to 60°C	0°C to 50°C
Storage Temperature	–40°C to 85°C	–10°C to 60°C
Relative Humidity	5 to 95% RH	5 to 90% RH

Table 7.1
Environmental conditions

Should personnel suspect that there could be problems with the environment having excessive dust, corrosive vapors, moisture or oil the best approach is to mount the computer system in an enclosure. This will provide the necessary protection for the processor. Special consideration may have to be given to such issues as vibration and it may be necessary to mount the computer in the enclosure on shock mountings to absorb some of the vibration. Ensure that the enclosure doors can be easily opened and that the heat is allowed to dissipate. Note that hot air rises and there can be build up of heat inside the top of the enclosure and a fan may be needed to circulate the air. Computer manufacturers have tables available providing data on the allowable ambient temperatures around their equipment. Generally computers do not immediately fail when the heat is excessive but have intermittent dropouts and suffer long-term damage.

The enclosure should be large enough to allow space to work on the system and to observe diagnostic lights/LEDs etc.

7.2.2   Earthing and shielding

Ensure that all hardware is securely earthed. The earth electrode is the central point for all electrical equipment and AC power within the facility. Use the maximum size copper wire (say, 8 AWG) for the earth.

Certain connections require shielded cables to reduce the effects of electrical noise. Ensure that only one end of the shield is earthed. As discussed in a previous chapter, earthing at both ends of a shielded cable should be avoided, as it will cause an earth loop in the cable.

7.2.3   Cabling

Full details on allowable distances for separating power and communications cables are given in Chapter 4. Some points to emphasize when installing communications cabling between the different computers and systems in the control room are listed below:

- Calculate the actual distance the cable is being run - i.e. both the horizontal and vertical distances. Select the shortest possible path away from sources of noise.
- Route the cables well away from potential sources of electrical interference, harsh chemicals, excessive heat, wet environments and sources of physical damage.
- Ensure that no one will walk or drive on the cable.

- Ensure that the cable is not put under undue tension (such as hanging between two points).
- Do not bend the cable excessively in the installation process.

7.2.4   Power connections

For installations near sources of electrical interference, an isolation transformer is a recommended approach. Note that the output devices being controlled should draw power from the original source of the voltage unless the secondary of the isolation transformer (which is supplying the computers) has been specifically rated for these additional devices.

Where the AC power source has variations, a constant voltage (CV) transformer can stabilize the voltage for short periods of time, thus minimizing shutdowns. It is worth noting here, that CV transformers are very sensitive to variations in main frequency and will not operate successfully with unstable mains frequency supplies.

For both the constant voltage transformer and the isolation transformer the operating frequency and the operating voltage should be carefully specified (e.g. 240 V AC +10% −15% or 50 Hz ± 2%).

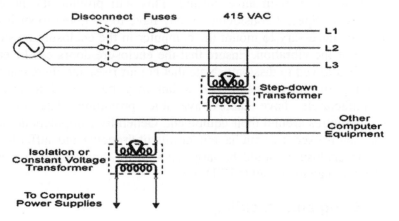

Figure 7.1
An isolation transformer

It is important to size transformers correctly:

- If the transformer is too small it will clip the peaks off the sine wave (due to saturation) resulting in a lower rms value of the voltage. The power supply could sense this as a low voltage and shutdown. The transformer may also overheat and burn out.
- Excessively large transformers do not provide as much isolation as a correctly sized transformer, due to higher capacitive coupling.

Useful techniques to reduce the electromagnetic interference and switching transients are given in Figure 7.2.

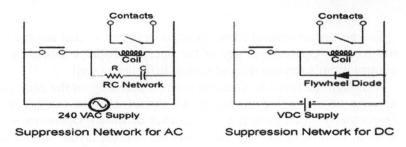

Figure 7.2
Techniques for reducing electromagnetic interference and surges

7.3 Ergonomic requirements

The main reason for considering ergonomic requirements is to improve the working environment of control room personnel. In the long run this should improve the productivity and reliability of the overall system.

The majority of tasks in a computer control room can be broken down into the following:

- Monitoring of the system
- Control adjustments
- Alarm/emergency procedures
- Staying awake

7.3.1 Typical control room layout

A typical layout is given in Figure 7.3.

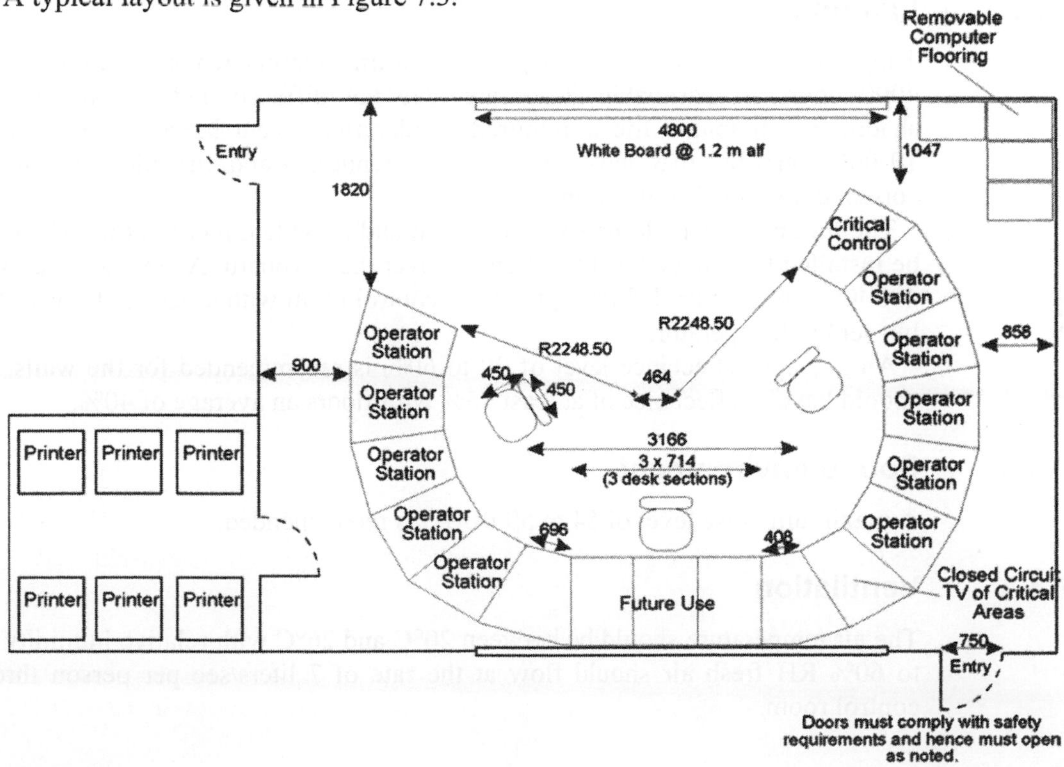

Figure 7.3
Typical layout of the computer control room

The horseshoe control room layout is designed so that anyone in the center can see all the screens. Operators at any of the operator displays should be able to view the entire control room's screens without undue difficulty as well.

Although the focus in a control room is normally on the equipment and computers, the amount of space for the operators should also be maximized to avoid congestion (particularly when there is a change over of shifts). Operators will spend a considerable amount of time in front of their consoles and the layout should ensure that the operator can see anyone coming into the control room and not have people peering over their shoulders.

Similar areas in the system that are being monitored should be situated close together to avoid unnecessary movement by the operators to see what is going on.

The voice communications system (either radio or telephone) should be situated as close as possible to the operators and for other persons entering the control room. For the control room indicated in the diagram at least three internal telephones should be provided for easy access (with frequently used numbers programmed into the system).

The amount of desk space should not be compromised. Space should be allowed for manuals and other items to be left on the desk without unnecessary clutter.

The printers for the system are situated in a separate room to isolate the operators from the associated (rather repetitive) noise. The associated inconvenience of having to walk to the printer room to view alarms can be minimized by providing on-screen alarm reports.

A separate meeting room should be provided to avoid holding meetings in the control room which are of no interest to the operator but which disrupt his work. The following specific issues should also be considered in the design of the computer control room.

7.3.2 Lighting

Tungsten halogen light sources produces warm lighting while the light life of 2000 to 4000 hours is reasonable. They are also not diffused and can produce significant shadowing. If longer life is required tubular fluorescent lamps have a life of 5000 to 10 000 hours but may have variable color rendering and variable apparent color if the correct color tube is not chosen.

The luminaries should be fixed overhead and provide direct lighting. Desk lighting can be installed to provide localized lighting over the keyboard. A general level of lighting of 400 lux is recommended throughout the control room with a personal level of 200 to 600 lux set by the operator.

An average reflectance level of 30 to 60% is recommended for the walls. The ceiling should have a reflectance of at least 75% with floors an average of 40%.

7.3.3 Sound environment

A maximum noise level of 54 to 59 dB(A) is recommended.

7.3.4 Ventilation

The air temperature should be between 20°C and 26°C with relative humidity range of 40 to 60% RH fresh air should flow at the rate of 7 liters/sec per person throughout the control room.

7.3.5 Colors of equipment

Colors for walls and equipment should have a matt finish (i.e. no shiny surfaces) to avoid irritating reflections from the operator displays. Strong contrasts in color should also be avoided to minimize glare.

Where the general light level is low (less than 300 lux) warm color schemes are more acceptable than those in which cold colors predominate. A pleasant color scheme can be achieved with warm colors backed up with cool secondary colors.

7.4 Design of the computer displays

The objective of this discussion is to provide a useful set of guidelines for the design of an effective operator display system. The approach should be to ensure that the displays are as easy to read and understand as possible. This reduces the decoding process in the human brain to a minimal level and maximizes the decision-making processes of the brain as per Figure 7.4. This ensures that the operator can react quickly and effectively without having to work out where the problem is.

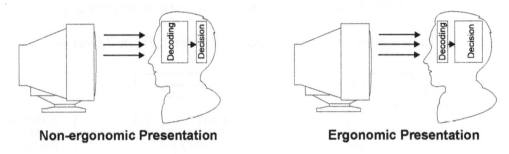

Non-ergonomic Presentation   **Ergonomic Presentation**

Figure 7.4
Ergonomic versus non-ergonomic representation of data

Typical hardware that is provided is:
- One or more operator displays (which may be of the touch type)
- Industrial (or Mylar) type keyboards which have audible or tactile feedback
- Operator panels consisting of highlighted keys to bring up predefined graphic displays
- Printers (one for alarms and one for reports)
- Alarm buzzers (or external sirens)

(A useful addition although possibly expensive option is a video copier for reproducing the operator screens in color.)
A useful configuration is as in Figure 7.5.

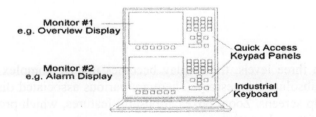

Figure 7.5
Configuration of the operator screens

Displays should appear within one second of the operator pressing the appropriate display key(s).

7.4.1 Operator displays and graphics

The organization of displays should be done in a clear and logical way to allow the operator to quickly and effectively identify the information of interest. The architecture of displays is to have a progressive decrease in scope of the displays and a progressive increase in detail as the operator looks for some specific information (and is proceeding down the hierarchy of displays).

Displays should be organized into three layers:

- The primary level which is an overview level and which should be reached directly from the function keys on the keypad.
- The secondary level, which consists of a number of displays, associated with that of the primary level. These should be able to be accessed directly from the primary level displays.
- The tertiary level, which gives more details on certain secondary level, displays.

The suggested layout of the displays is given in Figure 7.6 below.

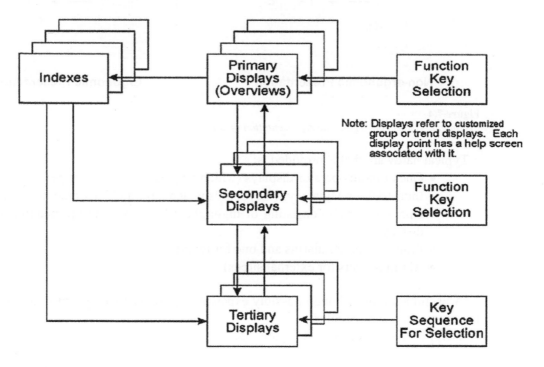

Figure 7.6
Display hierarchy

With more than three levels, the display becomes unduly complex and this should be avoided unless absolutely necessary. There are various associated displays such as trend displays and help screens. Zoom boxes are useful features, which provide more detail on a specific area of the schematic.

The various graphic screens that are available are:

- **Free form graphic screens**

  This is where the screen format can be created by the users, using whatever layout and symbols they can create. These are best constructed by the operators (with assistance from the engineer). They offer the designer complete flexibility in the layout of the information.

- **Operating group displays**

  Here a standard set of symbols is used to create displays as required. These provide the data in a standard presentation format.

- **Trend displays**

  These displays occupy part or the entire screen depending on the configuration. They provide trends on the data of analog values.

- **Alarm displays**

  These log the current alarms for the system.

It has been shown that operators consult and use overview type schematics at least ten times more often than secondary and tertiary schematics. It is thus imperative that as much effort as possible goes into the correct design of such displays. The operators should be consulted as much as possible in the design of these screens to make them as useful as possible.

Overview displays have to cover a large amount of the system and it is thus important to eliminate any part of the display, which does not convey information to the operator. This would mean that equipment outlines and flow lines are not put into overview schematics.

Secondary and tertiary displays are consulted less and probably require more information than that of the live updates. Outlines of equipment and text messages should be de-emphasized by using low intensity colors.

An icon should be designed to indicate clearly the area associated with a given schematic or operation. This allows the operator to quickly work out which area the current display is referring to.

7.4.2 Design of screens

There are generally two conflicting demands made on the design of screens:

- Reduce the complexity of the screen
- Try and keep all the displays associated with a given function to a minimum.

There are a number of effective ways in which to design appropriate operator screens. A few techniques to bear in mind are:

- **Signal to noise ratio**

  If the signal (on any display) is defined as the information the operator is looking for and noise is information the operator does not require, it is clear that the purpose of a particular screen should be carefully defined so as to maximize live data relating to the current function, and minimization of irrelevant data (and detailed graphic layouts of shape of vessels and piping etc).

- **Color and symbols**
  Color chart guidelines should be followed where possible. Composite symbols and flags should be devised for devices with different states such as on/off/tripped/unavailable etc, for a pump. Flow lines and tanks should be dark blue when inactive and red (for example) when active.

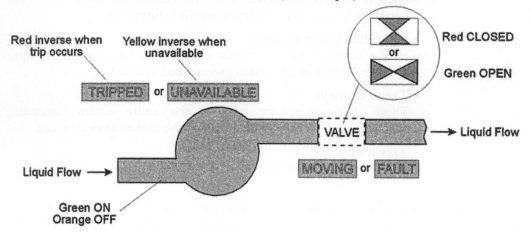

Figure 7.7
Composite pump symbol

Patterns should be used where possible to reduce the complexity and clutter of the display. Do not exceed 100 mm by 100 mm areas with patterns, as it is difficult for the operator to comprehend areas larger than this on one screen.

High priority areas on the screen should be highlighted with high intensity and bright colors. Dull unattractive colors should be used for unimportant items. Alternatively, unimportant items should be left off the screen.

Outlines of equipment items should be simple; a life-like representation is not needed and may in fact clutter the screen unnecessarily.

Outlines should be clearly differentiated from each other; even if this requires a certain amount of exaggeration.

Do not fill outlines up unless absolutely necessary (and then only with dull colors).

7.5 Alarming and reporting philosophies

Alarm processing is an important part of the operator station. Error codes identifying the faults are normally included with the description of the failed device.

No other part of the operator display has as much impact on the health of the plant (and that of the operator). The alarm function should be viewed as an integral part of the operator interface and not as a stand-alone feature. Figure 7.8 gives a view on the actions that occur on an alarm being activated.

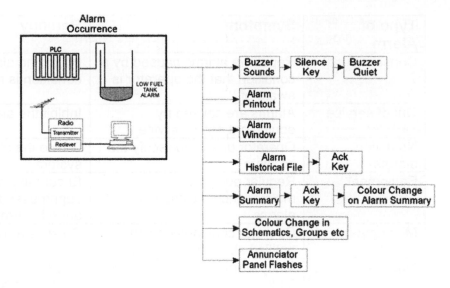

Figure 7.8
Alarm actions in an operator display

Another approach as opposed to the pure screen listing of alarms is to have an associated enunciator panel (situated next to the operator display) with illuminated pushbuttons. Each pushbutton would indicate the area from which the alarms originate and also when depressed would cause the appropriate schematic to appear on the operator display.

Only four alarm priorities should be implemented. These are:

- **High priority**
  Alarms that warn of dangerous conditions that could cause a shutdown of a major activity.

- **Medium priority**
  Alarms that should be acted on as quickly as possible; but will not cause a shutdown.

- **Low priority**
  Alarms that should be dealt with when time permits.

- **Event only**
  Statistical or technical information. No enunciator sounds for these.

The limiting of the number of types of alarms is to keep the system straightforward and with easy interpretation of the alarms. Higher priority alarms should be louder; lower pitched and have a higher pulse frequency than the lower priority alarms. Alarms are classified as unacknowledged (and flashing on the screen) until the operator acknowledges them via the keyboard. They then become an accepted alarm. One weakness in many alarm systems is the occurrence of trivial alarms, which irritate and confuse the operator. Typical trivial alarms are summarized in the table below:

Type of Alarm	Symptom	Remedy
Consequential	Repetitive alarms caused by a condition that the operator is aware of	Inhibit the alarm until the condition is remedied
Out of service	Alarms are caused by equipment not in service	Inhibit the alarms
No action alarms	Operator unable to rectify the problem	Delete the alarm from the system
Equipment changes	Regular equipment maintenance etc causes alarms	Ensure the alarms are suppressed for this period by added alarm logic
Minor event	Operator constantly being notified about trivial events	Delete alarm and replace with event recording
Multiple	Many alarms triggered by one fault	Use first up alarming to reduce the alarm information
Cycling	Signal close to alarm level moves the alarm in and out of alarm condition	Expand the range of signal before moving into alarm
Instrument drift	Drift of instrument causes alarm	Ensure there is tight control on the calibration of instruments

Table 7.2
List of trivial alarms

It is important to continuously audit, maintain, and improve on the alarm system through analysis and review with the operators on the performance of the system.
For every alarm the following should be documented:

- Type of alarm
- Alarmed tag
- Description of tag
- Reasons for alarm
- Relationship to related alarms (consequential relationships)
- Description of the logic in the generation of the alarm
- Possible causes of the alarm
- Action steps to take to remedy the alarm situation

Alarms should be able to be disabled provided the operator has the relevant key.
Suggested colors for alarms could be:

RED	HIGH PRIORITY
MAGENTA	MEDIUM PRIORITY
YELLOW	LOW

8

Troubleshooting and maintenance

8.1 Introduction

This section reviews certain methodologies that may be followed for effective troubleshooting and maintenance of a telemetry system from the digital or analog field input/output at the RTU to the computer facilities at the central site. The emphasis in this section is on the methodology to be used and reference will be made to sections throughout the manual to the details of the equipment. Basic troubleshooting of the various components of a telemetry system will be covered here. This includes:

- The RTU and component modules
- Associated equipment interfaced to the RTU (such as PLCs)
- Radio transceivers
- Antennas and antenna feeder systems
- The master station
- The central site computer facilities
- Maintenance aspects will be considered at the end of the section.

An overall diagram for the entire system is as follows:

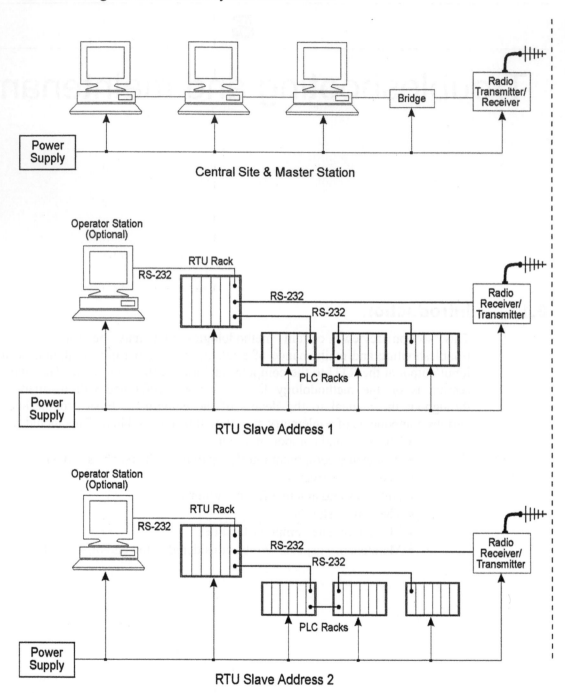

Figure 8.1
Troubleshooting the overall system

8.2   Troubleshooting the telemetry system

Ensure that components are not removed on line whilst the system is powered up unless the manufacturer specifically indicates that this is permissible. Damage to components and modules can occur when removing whilst the system is still powered up. Ensure that the antenna system is not disconnected from the system unless a dummy load has been installed, otherwise the radio power amplifier may be damaged.

8.2.1   The RTU and component modules

A typical procedure to follow when reviewing the operation of the telemetry system for faults (either for intermittent or outright failure) is:

- Confirm that the power supply module is healthy. Check the main fuse or circuit breaker of the equipment rack or unit if no power is evident.
- If the power supply is not operating, check that there is power to the power supply module. If there is power to the module then replace the power supply module.
- Check central processing (CPU) module that the run or healthy light is on. Switch the CPU module to run mode if not running.
- Check earthing connections for low resistance to earth or whether some other drive hardware (such as a variable speed drive) has been added to the system.
- If CPU module will not run, check the configuration program to see whether this is faulty or not. Reload the program if indications are that it has become corrupted. Check that the configuration of the system matches that of the hardware. Load up a simpler program that you know works, if the program is possibly defective. Back up defective program onto disk (for future analysis) and reset the memory.
- Cycle power to the RTU before the new program is loaded.
- Replace the CPU and retry the test.
- Check that the modem module is operating. Confirm that the modem is operational and that it is transmitting and receiving data by examining the transmit (TX) and receive (RX) lights on the front panel.
- If the modem is not operational, replace the modem module (or desktop unit).
- If the modem module is not operating correctly, perform the local and remote loopback tests as described in the modem section.
- Check each analog and digital input/output module for status such as healthy or run.
- Check for possible heating problems in the system cabinet.
- This could be due to failure of an air-conditioning or fan unit (if installed) or excessive ambient temperature.
- If a module indicates no power at all; check the fuse for that module.
- Replace each module if the indications are not healthy.

Check each individual module as follows:

Analog input modules

- Check that there is current or voltage being injected into the signal inputs.
- Check the fuse is installed.

- Check the scale and span and compare with the appropriate register tables for accuracy.
- Adjust scale and span either via software or via pots on the card.

Digital input module

- Check that there is current or voltage at the signal input to the module.
- Check the fuse for each input.
- Check earthing connections.

Interface from RTU to PLC (RS-232/RS-485)

- Check for the transmit/receive/run lights on the interface unit.
- Check interface data communications link.
- Check that the radio unit is operating (if used). The on light should be on and no fault lights should be on.
- If the on light is not on, check that there is DC power to the radio. If there is power, check that the fuse or circuit breaker is healthy.
- Check that all coaxial connectors are secured properly.
- Check that the PTT (transmit) light on the radio comes on when the telemetry unit feeds data into it for transmitting. Check that there is sufficient audio level into the transmitter.
- Check that the mute (receive) light on the radio comes on when the radio receives RF data. Check that there is sufficient audio level into the telemetry unit from the receiver.
- Check that the VSWR into the antenna is 1.5 or less.
- Check that the RF output power from the transmitter is as specified.
- Check that the antenna is aligned in the correct direction and with correct polarization.
- If the radio is still not working correctly, a radio test set will be required to check transmitter deviation, RF distortion, audio distortion, receiver sensitivity at 12 dB SINAD, transmitter and receiver frequency errors, transmitter/receiver isolation and transmitter spurious outputs.
- If a landline is to be used, then first ensure that the telemetry and modem equipment is operational. Then if the line is a:

Privately owned cable

- Check for end to end connectivity
- Measure noise level on line
- Check for crossed pairs
- Check MDF and IDF connections
- Check earthing

Switched telephone line

- Listen for dial tone
- Connect up a standard telephone and make a normal telephone call
- Listen for noise levels
- Call out the telephone company

Analog or digital data links

- Check the run, connect, transmit and receive lights on the modem
- Check the operations manual and the communications software
- Call out the telephone company

8.2.2   The master sites

The master sites will generally consist of a more complete telemetry unit and higher quality radio equipment. The same troubleshooting techniques can be applied to the master site as are used at the RTUs. Additional equipment would include links to other master sites, to the central site and computer control facilities.

The additional checks that will be appropriate include:

- Check that the link to the central site is operating correctly.
- If it is a radio link, carry out the check as discussed in the last section.
- If it is a microwave link, check that the transmit and receive lights are on.
- Check that the BER alarm light is not on.
- Check transmit power.
- Check receiver sensitivity.
- Check antenna alignment.
- Check connectors are secure and the cable or waveguide has not been damaged.
- Check individual multiplex cards for alarms and power fails.
- Check input/output levels from multiplex cards.
- Check for clean (noise free) healthy power supply to microwave equipment.
- Check link fade margin.
- As a last resort, carry out BER tests on each channel.
- If there is a master site computer, check that it can carry out all its required functions, i.e. monitoring of radio and RTU performance, status and alarms, etc.

The next section offers some hints as to how to troubleshoot a computer system.

8.2.3   The central site

The areas in which to troubleshoot problems here are quite varied as the master station consists of:

- The operator stations
- The software for the system
- The communications network for the operator stations

The radio and the antenna systems have already been covered in the previous section. However the operator stations, the communications network and the associated software will be covered in this section.

8.2.4   The operator station and software

There is not much that can be done here if a system fails or has intermittent problems except to systematically replace each connected unit to identify the faulty module. This would typically involve replacing the following units in turn:

- Operator terminal (normally a personal computer)
- Local area network card(s)
- Bridge unit to radio, microwave or landline system
- Printer connected to operator terminal

There are however a few problems which can be examined:

- **Operator terminal locks up intermittently**
  - Check the power supply to the system for possible electrical spikes or transients. This can be done with power analysis equipment (e.g. the dranetz) or by putting the entire system onto a battery supply.
  - Check for any new electrical drives or pieces of equipment, which have recently been added to the system (and which may add harmonics to the system).
  - Check the earthing cable connections that the impedance to earth is still to specifications (typically less than 1 ohm).
  - Do a software check on the hard disk of the operator terminal for possible corruption of software or failure of the disk. Backup the system, reformat the hard disk and reinstall the software on the disk.
  - Replace the motherboard on the operator terminal (this probably indicates that the operator terminal should be replaced with another system).
- **Throughput of the operator station and associated system drops off dramatically**
  - This manifests itself in slow updates of data on the operator terminal.
  - Check the system for errors being introduced on the data communications lines by electrical noise or earthing problems. The data communications system could be sending multiple messages due to errors introduced by electrical noise.
  - Check the local area network for potential overload due to excessive traffic. Reduce the traffic by reducing the amount of data being transferred or split the systems up into separate networks (using bridges).
  - Check the radio, microwave, landline, and antenna systems for possible introduction of noise and error problems.

8.3   Maintenance tasks

There are a number of tasks that should be regularly undertaken to ensure that the complete telemetry system remains operational and performs up to the required standard. Whether the maintenance tasks are to be undertaken internally by your company or outsourced to a maintenance contractor, it is vital that there is a comprehensive and planned approach to maintenance.

The following are a list of key ideas and requirements when setting up a maintenance program.

- Produce a complete and thorough inventory of every piece of equipment in the telemetry system. This should include all single items of equipment. It should be the responsibility of the maintenance personnel to keep this current on a computer database.
- Define goals for system and equipment item working availability. Depending on the size and quality of the system:
- System availability of between 99.5% and 99.9% over 12 months.

- Individual equipment item availability of between 95% and 98% over 12 months.
- Set up a direct line of communications from the maintenance personnel to the company representative in charge of the telemetry system. All problems and potential problems should be directly and immediately relayed to that person.
- Maintenance activities should be clearly defined in terms of daily, weekly, monthly and annual activities. These should include:
- **Daily**
  - Repair of any immediate system faults.
  - A check from the central site computer monitoring facilities that there are no system problems.
  - Continued repair of faulty equipment items.
- **Weekly**
  - Produce a short weekly report of the system performance and problems encountered.
  - Perform functional checks of each piece of equipment in the system to ensure it is operating.
- **Monthly**
  - Visit each RTU and master site and carry out a thorough visual check of each equipment item. Then give the site a functional check.
  - Clean each site of dust and vermin.
  - Carry out a number of telemetry operations in conjunction with the central and master sites.
- **Annual**
  - Carry out a complete and thorough audit of the entire system.
  - Test and measure the main operating parameters of every piece of equipment. This should include voltages, currents, frequencies, inputs, outputs, levels, noise, etc.
  - A comprehensive report of the status of the equipment should be produced at the end of the audit.
  - Every RTU and master site should have a separate maintenance logbook, which should be filled in every time a maintenance person visits the site.
  - Monthly and annual meetings should be carried out to discuss maintenance problems and establish future objectives.
  - A complete and up-to-date inventory of spare parts should be kept so that stocks can be replenished when they get low.
  - A complete and up-to-date database of radio frequency licenses should be kept so that they can be paid for on an annual basis.
  - If a contractor is carrying out the maintenance, always ensure that they have sufficient personnel to carry out the work (in particular during annual leave periods).
  - Make sure all changes to the system are fully documented and drawings are updated.

8.4   The maintenance unit system

In industry today there is an increasing tendency for companies to reduce permanent staff to cover only the core business of the company and to contract out peripheral activities to specialist maintenance companies.

There are facets to be considered in taking such a decision and one of these is the question of how to adjust the current value of a maintenance contract where the amount of equipment varies from time-to-time. The following case study is drawn from experience with a large iron ore mining company which took the decision to out-source communications maintenance.

To begin with there was a very wide range of equipment to be maintained. There were several hundred mobile radios installed in mining equipment, locomotives, railroad maintenance machines, re-claimer excavators, ordinary vehicles, and offices. There were many isolated radio repeater sites with a combination of wind and diesel generating plants, as well as radio, multiplex equipment, telemetry systems, batteries, and antenna systems. In addition, there were numerous other items of equipment (much of it not related to the communications system) that were added to the contract from time-to-time, as an ongoing service to the mining company.

The starting point in the establishment of the maintenance unit (MU) system was the setting up of a maintenance program based on the system of 'A', 'B' and 'C' levels of service which is widely used for vehicle maintenance. In this system, an 'A' service is a simple visual inspection and performance check which is performed weekly or monthly; a bit like tyres, oil and water for a vehicle. A 'B service is less frequent and involves a detailed inspection, several measured performance checks and perhaps the generation of some form of written report. A 'C' service would take place perhaps annually and involved a major inspection, particularly of mechanical components and a detailed performance check with a printed report which made recommendations to the company for the future service life of the equipment.

The next stage was the establishment of a complete list of all equipment under maintenance and the estimated time required to do the maintenance, and by correlating this data it was possible to come to a man-hours total for the task. To this, allowances for emergency fault clearance and supervision were added and from this total the number of technicians required to do the work was established.

The next step was to find a total cost and this involved the costs of wages, allowances, supervision, air fares, vehicle costs, housing and associated expenses and a profit margin for the contractor.

At this point, mobile radios were designated as the sort of lowest common denominator in the system and they were given a value of 1 MU. Then MU values were assigned to all other types of equipment under maintenance and these ranged from perhaps 5 MU for a radio link terminal to 50 MU for a diesel power plant located 200 km from the depot. One of the merits of the system was that it was possible to load the MU value of a piece of equipment in relation to the complexity, reliability and difficulty of access so that traveling time was taken into account.

From the above it was possible to total the number of MUs and to relate this to the total cost and so a unit value for the MU was obtained.

When the contractor was given additional equipment to look after this was tallied in a monthly report along with the purchase order number or delivery docket, etc and when equipment was written off or removed from service, this was included with the appropriate advice reference. In this way, the company had a monthly record of all equipment under maintenance with additions and withdrawals highlighted.

During contract negotiations, a formula for the variation in the value of the MU was developed and this took into account factors such as award variations, fuel prices, and CPI adjustments.

The end result was a monthly total of MUs, which was multiplied by an up-to-date cost per MU to produce a fully substantiated invoice cost.

The above system operated satisfactorily for several years. It was periodically reviewed by auditors on both sides to ensure that the MU cost truly reflected the contractors' costs and served both parties well. It may serve as a model for other similar maintenance agreements.

9

Specification of systems

9.1 Introduction

To successfully implement a quality telemetry system it is important to provide potential contractors with an unambiguous tender document. All companies will have a specific and well-established form of tender. This section describes the particular aspects of telemetry systems that need special attention in a tender document.

9.2 Common pitfalls

There is a vast array of telemetry equipment available on the market, servicing everything from small low integrity applications to large sophisticated systems. Before attempting to specify a system, it is best to visit and inspect as many different telemetry suppliers as possible. This will provide a good feeling for the levels of systems available and help evaluate the level of equipment suitable for the application.

It is important not to expect too much from equipment unless you are willing to pay for it. Over specifying a system, by asking for more than what is really needed can see a project fail because of blowouts in budgets. The author has seen many a bright-eyed young engineer's heart sink. If a particular feature is not required at present but may be in the future, do not specify it as a requirement initially but include a requirement that the system be easily expandable at any time to include the feature.

The other major pitfall is to not specify the complete system. Do not assume that a feature is included in the equipment because it says so in the manufacturer's literature. Write every trivial technical requirement into the specification. Before writing the specification, do a thorough survey of the application. Talk to all relevant personnel who will be involved in the implementation, management, and maintenance of the system. Carefully list every requirement in the tender document.

A further consideration is the choice of communications medium. Do not assume that general vendor information will necessarily be true for your application (they have vested interests). Do a careful budgetary analysis of different mediums. Ensure that the medium can provide the quality of data transmission required. Obtain budgetary prices and performance specifications of equipment and services from suppliers prior to tender to help evaluate the best medium. Where possible, visit a number of installations to evaluate how they are performing.

9.3 Standards

It is important to list all relevant local, national, and international standards and recommended practices that are applicable to the system to be installed. Standards should cover:

- Performance of electronic equipment
- Performance of communications equipment
- Installation practices
- Mechanical construction
- Environmental performance
- High and low voltage electrical requirements
- Health and safety requirements
- Rules and regulations that are applicable to working on site
- Production of drawings
- Production of documentation

A national organization in your country that takes responsibility for establishing a base of standards from which industry can work. They publish a significant number of national technical standards. They will generally also stock all important international standards. Lists of standards can be provided by them, from which references to relevant standards can be included in a tender document. Care should be taken not to include reference to standards that are not relevant to the project. Copies of standards can be obtained at libraries for checking.

9.4 Performance criteria

It is very important to specify the level of performance that is required from a newly installed system. This should include both equipment reliability and communications link availability. Realistic values of both complete system and individual equipment item reliability should be specified. The tenderers should be asked to provide MTBF figures for main items of equipment.

The communications link should be designed before writing the specification. In this way realistic values of availability can be determined and included as minimum availability criteria. Again, the tenderer should be asked to provide what their designed availability is for each link.

9.5 Testing

Testing of the telemetry system should be done at four stages:

1. Each individual equipment item as it is manufactured.
2. The complete system as mock up in the factory.
3. Once the system has been installed on site.
4. Final commissioning and acceptance.

It should be requested that all four stages be carried out and documented and the documentation forwarded for approval upon completion. It need only be necessary for the company representative to witness stages (2) and (4). Stage (3) is to ensure everything is working before commissioning and then normally, the same tests are carried out in stage (4). The tests that are carried out should check every parameter of the system and its equipment. They should be full and comprehensive and carried out with quality test equipment.

It is not necessary to list every test in the tender document but to state that it is at the discretion of the company to carry out whatever tests they feel appropriate.

It is best to specify that the contractor provide a list of tests to be carried out prior to testing begins at all four stages. It will then be possible to screen them and add to them if required.

It is always best to do too many tests than not enough as this will hopefully find any potential problems before they occur. With telemetry systems, be sure to test every permutation of digital and analog input and output.

9.6   Documentation

One of the great failures of many industrial and communication systems is the lack of good documentation. Poor documentation makes maintenance of installed systems very difficult. Therefore, the quality of documentation required should be clearly described in the tender document.

There should be (at least) two maintenance handbooks provided for each type of equipment. Major items should also have an operating manual provided.

An overall system handbook should also be provided. This should contain an extensive description of the system, including design, operating and performance data. It should contain an overall approach to operation and maintenance of the system.

Included in the system manual should be all system drawings, plus the procedures and results of the commissioning tests.

System drawings are probably the most important of all system documentation. At a minimum, the following should be requested from the contractor:

- Block and level diagrams
- Input/output logic diagrams
- Termination schedules including connector details
- Equipment interconnection diagrams of all sections of the system
- Rack elevations and equipment layouts
- Equipment room layout
- Mast and antenna layouts
- AC power wiring schematics
- DC power distribution and wiring schematics
- MDF, IDF and FDP schematics
- Jumpering records

9.7   Future trends in technology

9.7.1   Software based instrumentation

Since the late 1980s there has been considerable growth in the provision of software-based instrumentation. A software instrument can be considered to be a way of presenting the data of the instrument in an effective and understandable way on a PC-based operator screen, for example. The user is able to create the necessary front panel interfaces to suit their specific requirements. The front panel on the operator screen (of a PC for example) can have knobs, switches, graphs and strip charts. This allows the user to display the inputs from the instrumentation system and shows the outputs to control devices in the field.

The user can position, size, label and configure the instrument's data type, dimension and range easily and effectively. This is a very flexible way of displaying instrumentation information as the faceplate can easily be modified for the application.

The typical steps one goes through to generate a complete system are firstly to build the front panel of the instrument with the necessary information displayed. The second step is to construct in block diagram form the functional blocks that interpret the commands from the user to the instrument and possibly do some background analysis of information coming in. These are constructed from palette menus. The whole structure is then compiled to ensure that its execution speed is as high as possible.

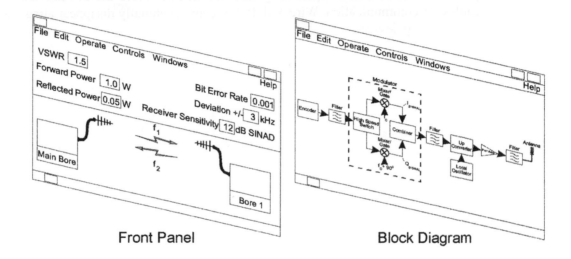

Front Panel   **Block Diagram**

Figure 9.1
Typical structure of a software based instrumentation system

The other obvious reason why software based instrumentation is becoming so popular is the lower cost and easy access to hardware than alternative approaches.

9.7.2   Future trends in SCADA systems

The future of SCADA systems is linked to 'company information services'. The overall trend in business is to move all data including SCADA on to HTML format. This will integrate the SCADA system into a complete company wide database. Hardware improvements in the past were overshadowed by better software. This will continue in the future. Companies will use www to access SCADA data from anywhere in the world. This will allow anyone in the company and even beyond to have access to SCADA derived data.

With the advent of the third party SCADA software package a complete interoperable system has been developed. Hopefully, the number of independent SCADA software companies will increase, unfortunately this has not been the case. It seems that the trend now is for large PLC companies to buy up independent SCADA software companies. This will bring into doubt the true interoperability of software packages such as CITECT, INTELLUTION and WONDERWARE. Without interoperable SCADA software systems, we will see a return to the bad old days of closed proprietary SCADA systems.

On the hardware side of SCADA, improvements in super smart sensors means that we will continue to see a reduction in the price and size of sensors. In addition, the functionality will increase at the same time. Fieldbus systems are becoming easier to set up and use. The move is to have a Fieldbus system that can be setup by almost anyone and is completely plug and play. This also will decrease repair cost and down time because everything will be modular. Super smart sensors allow the user to see problems, by way of the SCADA software, all the way down to the sensor level.

The move to a universal protocol is slow at best and probably will be overshadowed by the use of Ethernet as a common carrier for data. Tunneling as it's called will continue to grow because of the ease and low cost of Ethernet LAN systems. Systems of the future will use many protocols but each protocol will be imbedded into a 100 MHz Ethernet packet within an OSI compatible system. We will see more use of radio and fiber optic cables for communication. Wire will decrease and eventually disappear altogether, except for powering devices.

Appendix A
Glossary

ABM	*asynchronous balanced mode*
ACE	*asynchronous communications element* – similar to UART
ACK	*acknowledge* (ASCII – control F)
Active filter	active circuit devices (usually amplifiers), with passive circuit elements (resistors and capacitors) and which have characteristics that more closely match ideal filters than do passive filters
Active/passive device	device capable of supplying the current for the loop (active) or one that must draw its power from connected equipment (passive)
Address	a normally unique designator for location of data or the identity of a peripheral device, which allows each device on a single communications line to respond to its own message
AFC	*automatic frequency control* – the circuit in a radio receiver that keeps the carrier frequency centered in the passband of the filters and demodulators automatically
AGC	*automatic gain control* – the circuit in a radio that keeps the carrier gain at the proper level automatically
Algorithm	normally used as a basis for writing a computer program – this is a set of rules with a finite number of steps for solving a problem
Alias frequency	a false lower frequency component that appears in data reconstructed from original data acquired at an insufficient sampling rate (which is less than two (2) times the maximum frequency of the original data)
ALU	*arithmetic logic unit*
Amplitude modulation	a modulation technique (also referred to as AM or ASK) used to allow data to be transmitted across an analog network, such

as a switched telephone network – the amplitude of a single (carrier) frequency is varied or modulated between two levels one for binary 0 and one for binary 1

Amplitude flatness	a measure of how close to constant the gain of a circuit remains over a range of frequencies
Analog	a continuous real-time phenomena where the information values are represented in a variable and continuous waveform
ANSI	*American National Standards Institute* – the principal standards development body in the USA
Apogee	the point in an elliptical orbit that is farthest from earth
Application layer	the highest layer of the seven-layer ISO/OSI Reference Model structure, which contains all user or application programs
Arithmetic logic unit	the element(s) in a processing system that perform(s) the mathematical functions such as addition, subtraction, multiplication, division, inversion, AND, OR, NAND and NOR
ARQ	*automatic request for transmission* – a request by the receiver for the transmitter to retransmit a block or frame because of errors detected in the originally received message
AS	*Australian Standard*
ASCII	*American Standard Code for Information Interchange* – a universal standard for encoding alphanumeric characters into 7 or 8 binary bits, drawn up by ANSI to ensure compatibility between different computer systems
ASIC	*application specific integrated circuit*
ASK	*amplitude shift keying – see* Amplitude modulation
Asynchronous	communications where characters can be transmitted at an arbitrary, unsynchronized point in time and where the time intervals between transmitted characters may be of varying lengths – communication is controlled by start and stop bits at the beginning and end of each character
Attenuation	the decrease in the magnitude of strength (or power) of a signal – in cables, generally expressed in dB per unit length
Attenuator	a passive network that decreases the amplitude of a signal (without introducing any undesirable characteristics to the signals such as distortion)
Auto tracking antenna	a receiving antenna that moves in synchronism with the transmitting device which is moving (such as a vehicle being telemetered)
AWG	*American Wire Gauge*
Balanced circuit	a circuit so arranged that the impressed voltages on each conductor of the pair are equal in magnitude but opposite in polarity with respect to ground

Band reject	a circuit that rejects a defined frequency band of signals while passing all signals outside this frequency range (both lower than and higher than)
Band pass filter	a filter that allows only a fixed range of frequencies to pass through – all other frequencies outside this range (or band) are sharply reduced in magnitude
Bandwidth	the range of frequencies available expressed as the difference between the highest and lowest frequencies is expressed in hertz (or cycles per second)
Base address	a memory address that serves as the reference point. All other points are located by offsetting in relation to the base address
Base band	*base band* operation is the direct transmission of data over a transmission medium without the prior modulation on a high frequency carrier band
Base loading	an inductance situated near the bottom end of a vertical antenna to modify the electrical length – this aids in impedance matching
Baud	unit of signaling speed derived from the number of events per second (normally bits per second) – however if each event has more than one bit associated with it the baud rate and bits per second are not equal
Baudot	data transmission code in which five bits represent one character – sixty four (64) alphanumeric characters can be represented – this code is used in many teleprinter systems with one start bit and 1.42 stop bits added
BCC	*block check character* – error checking scheme with one check character; a good example being block sum check
BCD	*binary coded decimal* – a code used for representing decimal digits in a binary code
BEL	Bell (ASCII for Control-G)
Bell 212	an AT&T specification of full-duplex, asynchronous, or synchronous 1200 Baud data transmission for use on the public telephone network
BERT/BLERT	*bit error rate/block error rate testing* – an error checking technique that compares a received data pattern with a known transmitted data pattern to determine transmission line quality
Bifilar	two conducting elements used in parallel (such as two parallel wires wound on a coil form)
BIOS	basic input/output system
Bipolar	a signal range that includes both +ve and –ve values
Bit stuffing	*bit stuffing with zero bit insertion* – a technique used to allow pure binary data to be transmitted on a synchronous transmission line – each message block (frame) is encapsulated between two flags, which are special bit sequences, then if the message data contains a possibly similar sequence, an

additional (zero) bit is inserted into the data stream by the sender, and is subsequently removed by the receiving device – the transmission method is then said to be data transparent

BIT (Binary Digit)	derived from 'BInary DigiT', a one or zero condition in the binary system
Bits per sec (BPS)	unit of data transmission rate
Block sum check	this is used for the detection of errors when data is being transmitted – it comprises a set of binary digits (bits) which are the Modulo 2 sum of the individual characters or octets in a frame (block) or message
Bridge	a device to connect similar sub-networks without its own network address – used mostly to reduce the network load
Broadband	a communications channel that has greater bandwidth than a voice grade line and is potentially capable of greater transmission rates – opposite of base band – in wide band operation the data to be transmitted are first modulated on a high frequency carrier signal – they can then be simultaneously transmitted with other data modulated on a different carrier signal on the same transmission medium
Broadcast	a message on a bus intended for all devices, which requires no reply
BS	*backspace* (ASCII Control-H)
BS	*British Standard*
BSC	*Bisynchronous transmission* – a byte or character oriented communication protocol that has become the industry standard (created by IBM) – it uses a defined set of control characters for synchronized transmission of binary coded data between stations in a data communications system
Buffer	an intermediate temporary storage device used to compensate for a difference in data rate and data flow between two devices (also called a spooler for interfacing a computer and a printer)
Burst mode	a high speed data transfer in which the address of the data is sent followed by back-to-back data words while a physical signal is asserted
Bus	a data path shared by many devices with one or more conductors for transmitting signals, data, or power
Byte	a term referring to eight associated bits of information; sometimes called a 'character'
Capacitance (mutual)	the capacitance between two conductors with all other conductors, including shield, short circuited to the ground
Capacitance	storage of electrically separated charges between two plates having different potential – the value is proportional to the surface area of the plates and inversely proportional to the distance between them

Cascade	two or more electrical circuits in which the output of one is fed into the input of the next one
Cassegrain antenna	parabolic antenna that has a hyperbolic passive reflector situated at the focus of the parabola
CCITT	*Consultative Committee International Telegraph and Telephone* – an international association that sets world-wide standards (e.g. V.21, V.22, V.22bis) – now referred to as the International Telecommunications Union (ITU)
Cellular polyethylene	expanded or 'foam' polyethylene consisting of individual closed cells suspended in a polyethylene medium
Channel selector	in an FM discriminator the plug-in module which causes the device to select one of the channels and demodulate the subcarrier to recover data
Character	letter, numeral, punctuation, control figure or any other symbol contained in a message
Characteristic impedance	the impedance that, when connected to the output terminals of a transmission line of any length, makes the line appear infinitely long – the ratio of voltage to current at every point along a transmission line on which there are no standing waves
Clock	the source(s) of timing signals for sequencing electronic events e.g. synchronous data transfer
Closed loop	a signal path that has a forward route for the signal, a feedback network for the signal and a summing point
CMRR	*common mode rejection ratio*
CMV	*common mode voltage*
CNR	*carrier to noise ratio* – an indication of the quality of the modulated signal
Common mode signal	the common voltage to the two parts of a differential signal applied to a balanced circuit
Common carrier	a private data communications utility company that furnishes communications services to the general public
Commutator	a device used to effect time-division multiplexing by repetitive sequential switching
Composite link	the line or circuit connecting a pair of multiplexers or concentrators; the circuit carrying multiplexed data
Conical scan antenna	an automatic tracking antenna system in which the beam is steered in a circular path so that it forms a cone
Contention	the facility provided by the dial network or a data PABX, which allows multiple terminals to compete on a first come, first served basis for a smaller number of computer posts
Correlator	a device, which compares two signals and indicates the similarity between the two signals
CPU	*central processing unit*
CR	*carriage return* (ASCII control-M)

CRC	*cyclic redundancy check* – an error-checking mechanism using a polynomial algorithm based on the content of a message frame at the transmitter and included in a field appended to the frame – at the receiver, it is then compared with the result of the calculation that is performed by the receiver – also referred to as CRC-16
Cross talk	a situation where a signal from a communications channel interferes with an associated channel's signals
Crossed planning	wiring configuration that allows two DTE or DCE devices to communicate – essentially it involves connecting pin 2 to pin 3 of the two devices
Crossover	in communications, a conductor which runs through the cable and connects to a different pin number at each end
CSMA/CD	*carrier sense multiple access/collision detection* – when two situations transmit at the same time on a local area network, they both cease transmission and signal that a collision has occurred – each then tries again after waiting for a predetermined time period
Current loop	communication method that allows data to be transmitted over a longer distance with a higher noise immunity level than with the standard EIA-232-C voltage method – a mark (a binary 1) is represented by current of 20 mA and a space (or binary 0) is represented by the absence of current
Data reduction	the process of analyzing a large quantity of data in order to extract some statistical summary of the underlying parameters
Data integrity	a performance measure based on the rate of undetected errors
Data link layer	this corresponds to layer 2 of the ISO Reference Model for open systems interconnection – it is concerned with the reliable transfer of data (no residual transmission errors) across the data link being used
Datagram	a type of service offered on a packet-switched data network – it is a self-contained packet of information that is sent through the network with minimum protocol overheads
dBi	a unit that is used to represent the gain of an antenna compared to the gain of an isotropic radiator
dBm	a signal level that is compared to a 1 mW reference
dBmV	a signal amplitude that is compared to a 1 mV reference
dBW	a signal amplitude that is compared to a 1 watt reference
DCE	*data communications equipment* – devices that provide the functions required to establish, maintain, and terminate a data transmission connection – normally refers to a modem
Decibel (dB)	a logarithmic measure of the ratio of two signal levels where $dB = 20\log_{10}V_1/V_2$ or where $dB = 10\log_{10}P1/P2$ and where V

	refers to voltage or P refers to power – note that it has no units of measure
Decoder	a device that converts a combination of signals into a single signal representing that combination
Decommutator	equipment for the demultiplexing of commutated signals
Default	a value or setup condition assigned, which is automatically assumed for the system unless otherwise explicitly specified
Delay distortion	distortion of a signal caused by the frequency components making up the signal having different propagation velocities across a transmission medium
DES	*data encryption standard*
Dielectric constant (E)	the ratio of the capacitance using the material in question as the dielectric, to the capacitance resulting when the material is replaced by air
Digital	a signal which has definite states (normally two)
DIN	*Deutsches Institut Für Normierung*
DIP	acronym for dual in line package referring to integrated circuits and switches
Diplexing	a device used to allow simultaneous reception or transmission of two signals on a common antenna
Direct memory access	a technique of transferring data between the computer memory and a device on the computer bus without the intervention of the microprocessor – also abbreviated to DMA
Discriminator	hardware device to demodulate a frequency modulated carrier or subcarrier to produce analog data
Dish	a concave antenna reflector for use at VHF or higher frequencies
Dish antenna	an antenna in which a parabolic dish acts a reflector to increase the gain of the antenna
Diversity reception	two or more radio receivers connected to different antennas to improve signal quality by using two different radio signals to transfer the information
DLE	*data link escape* (ASCII character)
DNA	distributed network architecture
Doppler	the change in observed frequency of a signal caused by the emitting device moving with respect to the observing device
Downlink	the path from a satellite to an earth station
DPI	*dots per inch*
DPLL	*digital phase locked loop*
DR	*dynamic range* – the ratio of the full-scale range (FSR) of a data converter to the smallest difference it can resolve – $DR = 2n$ where n is the resolution in bits

Drift	a (normally gradual) change in a component's characteristics over time
Driver software	a program that acts as the interface between a higher level coding structure and the lower level hardware/firmware component of a computer
DSP	*digital signal processing*
DSR	*data set ready* – an EIA-232 modem interface control signal, which indicates that the terminal is ready for transmission
DTE	*data terminal equipment* – devices acting as data source, data sink, or both
Duplex	the ability to send and receive data simultaneously over the same communications line
Dynamic range	the difference in decibels between the overload or maximum and minimum discernible signal level in a system
EBCDIC	*extended binary coded decimal interchange code* – an eight bit character code used primarily in IBM equipment – the code allows for 256 different bit patterns
EDAC	*error detection and correction*
EIA	*Electronic Industries Association* – a standards organization in the USA specializing in the electrical and functional characteristics of interface equipment
EIA-232-C	interface between DTE and DCE, employing serial binary data exchange – typical maximum specifications are 15 m at 19 200 baud
EIA-422	interface between DTE and DCE employing the electrical characteristics of balanced voltage interface circuits
EIA-423	interface between DTE and DCE, employing the electrical characteristics of unbalanced voltage digital interface circuits
EIA-449	general purpose 37-pin and 9-pin interface for DCE and DTE employing serial binary interchange
EIA-485	the recommended standard of the EIA that specifies the electrical characteristics of drivers and receivers for use in balanced digital multipoint systems
EIRP	*effective isotropic radiated power* – the effective power radiated from a transmitting antenna when an isotropic radiator is used to determine the gain of the antenna
EISA	*Enhanced Industry Standard Architecture*
EMI/RFI	*electromagnetic interference/radio frequency interference* – 'background noise' that could modify or destroy data transmission
EMS	*expanded memory specification*

Emulation	the imitation of a computer system performed by a combination of hardware and software that allows programs to run between incompatible systems
Enabling	the activation of a function of a device by a defined signal
Encoder	a circuit which changes a given signal into a coded combination for optimum transmission of the signal
ENQ	*enquiry* (ASCII Control-E)
EOT	*end of transmission* (ASCII Control-D)
EPROM	*erasable programmable read only memory* – non-volatile semiconductor memory that is erasable in an ultraviolet light and re-programmable
Equalizer	the device, which compensates for the unequal gain characteristic of the signal received
Error rate	the ratio of the average number of bits that will be corrupted to the total number of bits that are transmitted for a data link or system
ESC	*escape* (ASCII character)
ESD	*electrostatic discharge*
Ethernet	name of a widely used LAN, based on the CSMA/CD bus access method (IEEE 802.3) – Ethernet is the basis of the TOP bus topology
ETX	*end of text* (ASCII control-C)
Even parity	a data verification method normally implemented in hardware in which each character must have an even number of 'ON' bits
Farad	unit of capacitance whereby a charge of one coulomb produces a one volt potential difference
Faraday rotation	rotation of the plane of polarization of an electromagnetic wave when traveling through a magnetic field
FCC	*Federal Communications Commission*
FCS	*frame check sequence* – a general term given to the additional bits appended to a transmitted frame or message by the source to enable the receiver to detect possible transmission errors
FDM	*frequency division multiplexer* – a device that divides the available transmission frequency range in narrower bands, each of which is used for a separate channel
FDM	*frequency division multiplexing* – the combining of one or more signals (each occupying a defined non-overlapping frequency band) into one signal
Feedback	a part of the output signal being fed back to the input of the amplifier circuit
FIFO	*first in, first out*

Filled cable	a telephone cable construction in which the cable core is filled with a material that will prevent moisture from entering or passing along the cable
FIP	*factory instrumentation protocol*
Firmware	a computer program or software stored permanently in PROM or ROM or semi-permanently in EPROM
Flame retardancy	the ability of a material not to propagate flame once the flame source is removed
Floating	an electrical circuit that is above the earth potential
Flow control	the procedure for regulating the flow of data between two devices preventing the loss of data once a device's buffer has reached its capacity
Frame	the unit of information transferred across a data link – typically, there are control frames for link management and information frames for the transfer of message data
Frequency domain	the displaying of electrical quantities versus frequency
Frequency modulation	a modulation technique (abbreviated to FM) used to allow data to be transmitted across an analog network where the frequency is varied between two levels – one for binary '0' and one for binary '1' – also known as *frequency shift keying* (or FSK)
Frequency	refers to the number of cycles per second
Full-duplex	simultaneous two way independent transmission in both directions (4 wire) – *see* Duplex
G	*Giga* (metric system prefix – 10^9)
Gain of antenna	the difference in signal strengths between a given antenna and a reference isotropic antenna
Gateway	a device to connect two different networks, which translates the different protocols
Geostationary	a special earth orbit that allows a satellite to remain in a fixed position above the equator
Geosynchronous	any earth orbit in which the time required for one revolution of a satellite is an integral portion of a sidereal day
GPIB	*general purpose interface bus* – an interface standard used for parallel data communication, usually used for controlling electronic instruments from a computer – also designated IEEE-488 standard
Ground	an electrically neutral circuit having the same potential as the earth – a reference point for an electrical system also intended for safety purposes
Half duplex	transmissions in either direction, but not simultaneously

Half power point	the point in a power versus frequency curve which is half the power level of the peak power (also called 3 dB point)
Hamming distance	a measure of the effectiveness of error checking – the higher the hamming distance (HD) index, the safer is the data transmission
Handshaking	exchange of predetermined signals between two devices establishing a connection
Harmonic distortion	distortion caused by the presence of harmonics in the desired signal
HDLC	*high level data link control* – the international standard communication protocol defined by ISO to control the exchange of data across either a point-to-point data link or a multidrop data link
Hertz (Hz)	a term replacing cycles per second as a unit of frequency
Hex	*hexadecimal*
HF	*high frequency*
High pass	generally referring to filters which allow signals above a specified frequency to pass but attenuate signals below this specified frequency
Horn	a moderate-gain wide-beamwidth antenna
Host	this is normally a computer belonging to a user that contains (hosts) the communication hardware and software necessary to connect the computer to a data communications network
I/O address	a method that allows the CPU to distinguish between different boards in a system – all boards must have different addresses
IA5	*International Alphabet number 5*
IEC	*International Electrotechnical Commission*
IEE	*Institution of Electrical Engineers*
IEEE	*Institute of Electrical and Electronic Engineers* – an American based international professional society that issues its own standards and is a member of ANSI and ISO
Impedance	the total opposition that a circuit offers to the flow of alternating current or any other varying current at a particular frequency – it is a combination of resistance R and reactance X, measured in ohms
Inductance	the property of a circuit or circuit element that opposes a change in current flow, thus causing current changes to lag behind voltage changes – it is measured in henrys
Insulation resistance (IR)	that resistance offered by insulation to an impressed DC voltage, tending to produce a leakage current though the insulation

Interface	a shared boundary defined by common physical interconnection characteristics, signal characteristics, and measurement of interchanged signals
Interrupt handler	the section of the program that performs the necessary operation to service an interrupt when it occurs
Interrupt	an external event indicating that the CPU should suspend its current task to service a designated activity
IP	*Internet protocol*
ISA	*Industry Standard Architecture* (for IBM personal computers)
ISB	*intrinsically safe barrier*
ISDN	*integrated services digital network* – the new generation of world-wide telecommunications network, that utilizes digital techniques for both transmission and switching – it supports both voice and data communications
ISO	*International Standards Organization*
Isotropic antenna	a reference antenna that radiates energy in all directions from a point source
ISR	*interrupt service routine* – see Interrupt handler
ITU	*International Telecommunications Union*
Jumper	a wire connecting one or more pins on the one end of a cable only
k (kilo)	this is 2^{10} or 1024 in computer terminology, e.g. 1 kB = 1024 bytes
LAN	*local area network* – a data communications system confined to a limited geographic area typically about 10 kms with moderate to high data rates (100 kbps to 100 Mbps) – some type of switching technology is used, but common carrier circuits are not used
LCD	*liquid crystal display* – a low power display system used on many laptops and other digital equipment
LDM	*limited distance modem* – a signal converter which conditions and boosts a digital signal so that it may be transmitted further than a standard EIA-232 signal
Leased (or private) line	a private telephone line without inter-exchange switching arrangements
LED	*light emitting diode* – a semi-conductor light source that emits visible light or infrared radiation
LF	*line feed* (ASCII Control-J)
Line turnaround	the reversing of transmission direction from transmitter to receiver or vice versa when a half duplex circuit is used
Line driver	a signal converter that conditions a signal to ensure reliable transmission over an extended distance

Linearity	a relationship where the output is directly proportional to the input
Link layer	layer 'two' of the ISO/OSI reference model – also known as the data link layer
Listener	a device on the GPIB bus that receives information from the bus
LLC	*logical link control* (IEEE 802)
Loaded line	a telephone line equipped with loading coils to add inductance in order to minimize amplitude distortion
Long wire	a horizontal wire antenna that is one wavelength or greater in size
Loop resistance	the measured resistance of two conductors forming a circuit
Loopback	type of diagnostic test in which the transmitted signal is returned on the sending device after passing through all, or a portion of, a data communication link or network – a loopback test permits the comparison of a returned signal with the transmitted signal
Low pass	generally referring to filters which allow signals below a specified frequency to pass but attenuated signals above this specified frequency are blocked
m	*meter* – metric system unit for length
M	*mega* – metric system prefix for 10^6
MAC	*media access control* (IEEE-802)
Manchester encoding	digital technique (specified for the IEEE-802.3 Ethernet base band network standard) in which each bit period is divided into two complementary halves; a negative to positive voltage transition in the middle of the bit period designates a binary '1', whilst a positive to negative transition represents a '0' – the encoding technique also allows the receiving device to recover the transmitted clock from the incoming data stream (self clocking)
Mark	this is equivalent to a binary '1'
Master/slave	bus access method whereby the right to transmit is assigned to one device only, the master, and all the other devices, the slaves may only transmit when requested
Master oscillator	The primary oscillator for controlling a transmitter or receiver frequency. The various types are: Variable Frequency Oscillator (VFO); Variable Crystal Oscillator (VXO); Permeability Tuned Oscillator (PTO); Phase Locked Loop (PLL); Linear Master Oscillator (LMO) or frequency synthesizer
Microwave	AC signals having frequencies of 1 GHz or more
MIPS	*million instructions per second*

Modem	*MODulator-DEModulator* – a device used to convert serial digital data from a transmitting terminal to a signal suitable for transmission over a telephone channel or to reconvert the transmitted signal to serial digital data for the receiving terminal
Modem eliminator	a device used to connect a local terminal and a computer port in lieu of the pair of modems to which they would ordinarily connect, allow DTE to DTE data and control signal connections otherwise not easily achieved by standard cables or connections
Modulation index	the ratio of the frequency deviation of the modulated wave to the frequency of the modulating signal
MOS	*metal oxide semiconductor*
MOV	*metal oxide varistor*
MTBF	*mean time between failures*
MTTR	*mean time to repair*
Multidrop	a single communication line or bus used to connect three or more points
Multiplexer (MUX)	a device used for division of a communication link into two or more channels either by using frequency division or time division
NAK	*negative acknowledge* (ASCII Control-U)
Narrowband	a device that can only operate over a narrow band of frequencies
Network	an interconnected group of nodes or stations
Network architecture	a set of design principles including the organization of functions and the description of data formats and procedures used as the basis for the design and implementation of a network (ISO)
Network layer	layer 3 in the ISO/OSI reference model, the logical network entity that services the transport layer responsible for ensuring that data passed to it from the transport layer is routed and delivered throughout the network
Network topology	the physical and logical relationship of nodes in a network; the schematic arrangement of the links and nodes of a network typically in the form of a star, ring, tree or bus topology
NMRR	*normal mode rejection ratio*
Node	a point of interconnection to a network
Noise	a term given to the extraneous electrical signals that may be generated or picked up in a transmission line. If the noise signal is large compared with the data carrying signal, the latter may be corrupted resulting in transmission errors
Non-linearity	a type of error in which the output from a device does not relate to the input in a linear manner

NRZ	*non return to zero* – pulses in alternating directions for successive 1 bits but no change from existing signal voltage for 0 bits
NRZI	*non return to zero inverted*
Null modem	a device that connects two DTE devices directly by emulating the physical connections of a DCE device
Nyquist sample theorem	in order to recover all the information about a specified signal it must be sampled at least at twice the maximum frequency component of the specified signal
Ohm (W)	unit of resistance such that a constant current of one ampere produces a potential difference of one volt across a conductor
Optical isolation	two networks with no electrical continuity in their connection because an optoelectronic transmitter and receiver has been used
OSI	*open systems interconnection*
Packet	a group of bits (including data and call control signals) transmitted as a whole on a packet switching network – usually smaller than a transmission block
PAD	*packet access device* – an interface between a terminal or computer and a packet switching network
Parallel transmission	the transmission model where a number of bits is sent simultaneously over separate parallel lines – usually unidirectional such as the Centronics interface for a printer
Parametric amplifier	an inverting parametric device for amplifying a signal without frequency translation from input to output
Parasitic	undesirable electrical parameter in a circuit such as oscillations or capacitance
Parity check	the addition of non information bits that make up a transmission block to ensure that the total number of bits is always even (even parity) or odd (odd parity) – used to detect transmission errors but rapidly losing popularity because of its weakness in detecting errors
Parity bit	a bit that is set to a '0' or '1' to ensure that the total number of 1 bits in the data field is even or odd
Passive filter	a circuit using only passive electronic components such as resistors, capacitors and inductors
Path loss	the signal loss between transmitting and receiving antennas
PBX	*private branch exchange*
PCM	*pulse code modulation* – the sampling of a signal and encoding the amplitude of each sample into a series of uniform pulses
PEP	*peak envelope power* – maximum amplitude that can be achieved with any combination of signals
Perigee	the point in an elliptical orbit that is closest to earth

Peripherals	the input/output and data storage devices attached to a computer e.g. disk drives, printers, keyboards, display, communication boards, etc
Phase shift keying	a modulation technique (also referred to as PSK) used to convert binary data into an analog form comprising a single sinusoidal frequency signal whose phase varies according to the data being transmitted
Phase modulation	The sine wave or carrier has its phase charged in accordance with the information to be transmitted.
Physical layer	layer 1 of the ISO/OSI reference model, concerned with the electrical and mechanical specifications of the network termination equipment
PLC	programmable logic controller
PLL	phase locked loop
Point-to-point	a connection between only two items of equipment
Polar orbit	the path followed when the orbital plane includes the north and south poles
Polarization	the direction of an electric field radiated from an antenna
Polling	a means of controlling devices on a multipoint line – a controller will query devices for a response
Polyethylene	a family of insulators derived from the polymerization of ethylene gas and characterized by outstanding electrical properties, including high IR, low dielectric constant, and low dielectric loss across the frequency spectrum
Polyvinyl chloride (PVC)	a general purpose family of insulations whose basic constituent is polyvinyl chloride or its copolymer with vinyl acetate – plasticisers, stabilizers, pigments and fillers are added to improve mechanical and/or electrical properties of this material
Port	a place of access to a device or network, used for input/output of digital and analog signals
Presentation layer	layer 6 of the ISO/OSI reference model, concerned with negotiation of a suitable transfer syntax for use during an application – if this is different from the local syntax, the translation to/from this syntax
Protocol	a formal set of conventions governing the formatting, control procedures and relative timing of message exchange between two communicating systems
PSDN	*public switched data network* – any switching data communications system, such as Telex and public telephone networks, which provides circuit switching to many customers
PSTN	*public switched telephone network* – this is the term used to describe the (analog) public telephone network
PTT	*Post, Telephone, and Telecommunications Authority*

QAM	*quadrature amplitude modulation*
QPSK	*quadrature phase shift keying*
Quagi	an antenna consisting of both full wavelength loops (quad) and Yagi elements
R/W	*read/write*
RAM	*random access memory* – semiconductor read/write volatile memory – data is lost if the power is turned off
Reactance	the opposition offered to the flow of alternating current by inductance or capacitance of a component or circuit
Repeater	an amplifier, which regenerates the signal and thus expands the network
Resistance	the ratio of voltage to electrical current for a given circuit measured in ohms
Response time	the elapsed time between the generation of the last character of a message at a terminal and the receipt of the first character of the reply – it includes terminal delay and network delay
RF	*radio frequency*
RFI	*radio frequency interference*
Ring	network topology commonly used for interconnection of communities of digital devices distributed over a localized area, e.g. a factory or office block – each device is connected to its nearest neighbors until all the devices are connected in a closed loop or ring – data are transmitted in one direction only – as each message circulates around the ring, it is read by each device connected in the ring
Ringing	an undesirable oscillation or pulsating current
Rise time	the time required for a waveform to reach a specified value from some smaller value
RMS	*root mean square*
ROM	*read only memory* – computer memory in which data can be routinely read but written to only once, using special means when the ROM is manufactured – a ROM is used for storing data or programs on a permanent basis
Router	a linking device between network segments which may differ in Layers 1, 2a and 2b of the ISO/OSI reference model
RS	*recommended standard* (e.g. RS-232-C). Newer designations use the prefix EIA (e.g. EIA-RS-232-C or just EIA-232-C)
RTU	*remote terminal unit* – terminal unit situated remotely from the main control system
SAA	*Standards Association of Australia*
SAP	*service access point*
SDLC	*synchronous data link control* – IBM standard protocol superseding the bisynchronous standard

Selectivity	a measure of the performance of a circuit in distinguishing the desired signal from those at other frequencies
Serial transmission	the most common transmission mode in which information bits are sent sequentially on a single data channel
Session layer	layer 5 of the ISO/OSI reference model, concerned with the establishment of a logical connection between two application entities and with controlling the dialogue (message exchange) between them
Short haul modem	a signal converter, which conditions a digital signal for transmission over DC continuous private line metallic circuits, without interfering with adjacent pairs of wires in the same telephone cables
Sidebands	the frequency components which are generated when a carrier is frequency-modulated
Upconverter	a device used to translate a modulated signal to a higher band of frequencies
Sidereal day	the period of an earth's rotation with respect to the stars
Signal to noise ratio	the ratio of signal strength to the level of noise
Simplex transmissions	data transmission in one direction only
Slew rate	this is defined as the rate at which the voltage changes from one value to another
SNA	*systems network architecture*
SNR	*signal to noise ratio*
SOH	*start of header* (ASCII Control-A)
Space	*absence of signal* – this is equivalent to a binary 0
Spark test	a test designed to locate imperfections (usually pin-holes) in the insulation of a wire or cable by application of a voltage for a very short period of time while the wire is being drawn through the electrode field
Spectral purity	the relative quality of a signal measured by the absence of harmonics, spurious signals and noise
Standing wave ratio	the ratio of the maximum to minimum voltage (or current) on a transmission line at least a quarter-wavelength long. (VSWR refers to voltage standing wave ratio)
Star	a type of network topology in which there is a central node that performs all switching (and hence routing) functions
Statistical multiplexer	a device used to enable a number of lower bit rate devices, normally situated in the same location, to share a single, higher bit rate transmission line – the devices usually have human operators and hence data are transmitted on the shared line on a statistical basis rather than, as is the case with a basic multiplexer, on a pre-allocated basis – it thus endeavors to exploit the fact that each device operates at a much lower mean rate than its maximum rate

STP	*shielded twisted pair*
Straight through pinning	EIA-232 and EIA-422 configuration that match DTE to DCE, pin for pin (pin 1 with pin 1, pin 2 with pin 2,etc)
STX	*start of text* (ASCII Control-B)
Subharmonic	a frequency that is a integral submultiple of a reference frequency
Switched line	a communication link for which the physical path may vary with each usage, such as the public telephone network
Synchronization	the co-ordination of the activities of several circuit elements
Synchronous transmission	transmission in which data bits are sent at a fixed rate, with the transmitter and receiver synchronized – synchronized transmission eliminates the need for start and stop bits
Talker	a device on the GPIB bus that simply sends information onto the bus without actually controlling the bus
Tank	a circuit comprising inductance and capacitance, which can store electrical energy over a finite band of frequencies.
TCP	*transmission control protocol*
TDM	*time division multiplexer* – a device that accepts multiple channels on a single transmission line by connecting terminals, one at a time, at regular intervals, interleaving bits (bit TDM) or characters (Character TDM) from each terminal
Telegram	in general a data block that is transmitted on the network – usually comprises address, information, and check characters
Temperature rating	the maximum, and minimum temperature at which an insulating material may be used in continuous operation without loss of its basic properties
TIA	*Telecommunications Industry Association*
Time sharing	a method of computer operation that allows several interactive terminals to use one computer
Time division multiplexing	the process of transmitting multiple signals over a single channel by taking samples of each signal in a repetitive time sequenced fashion
Time domain	the display of electrical quantities versus time
Token ring	Collision free, deterministic bus access method as per IEEE-802.2 ring topology
TOP	*Technical Office Protocol* – a user association in USA, which is primarily concerned with open communications in offices
Topology	physical configuration of network nodes, e.g. bus, ring, star, tree
Transceiver	a combination of transmitter and receiver

Transceiver	*transmitter/receiver* – network access point for IEEE-803.2 networks
Transient	an abrupt change in voltage of short duration
Transmission line	one or more conductors used to convey electrical energy from one point to another
Transport layer	layer 4 of the ISO/OSI reference model, concerned with providing a network independent reliable message interchange service to the application oriented layers (layers 5 through 7)
Trunk	a single circuit between two points, both of which are switching centers or individual distribution points – a trunk usually handles many channels simultaneously
Twisted pair	a data transmission medium, consisting of two insulated copper wires twisted together – this improves its immunity to interference from nearby electrical sources that may corrupt the transmitted signal
UART	*universal asynchronous receiver/transmitter* – an electronic circuit that translates the data format between a parallel representation, within a computer, and the serial method of transmitting data over a communications line
UHF	*ultra high frequency*
Unbalanced circuit	a transmission line in which voltages on the two conductors are unequal with respect to ground e.g. a coaxial cable
Unloaded line	a line with no loaded coils that reduce line loss at audio frequencies
Uplink	the path from an earth station to a satellite
USRT	*universal synchronous receiver/transmitter – see* UART.
UTP	*unshielded twisted pair*
V.35	CCITT standard governing the transmission at 48 kbps over 60 to 108 kHz group band circuits
VCO	*voltage controlled oscillator* – uses variable DC applied to tuning diodes to change their junction capacitances – this results in the output frequency being dependent on the input voltage
Velocity of propagation	the speed of an electrical signal down a length of cable compared to speed in free space expressed as a percentage
VHF	*very high frequency*
Volatile memory	an electronic storage medium that loses all data when power is removed
Voltage rating	the highest voltage that may be continuously applied to a wire in conformance with standards of specifications
VSD	*variable speed drive*

VT	*virtual terminal*
WAN	*wide area network*
Waveguide	a hollow conducting tube used to convey microwave energy
Word	the standard number of bits that a processor or memory manipulates at one time – typically, a word has 16 bits
X.21	CCITT standard governing interface between DTE and DCE devices for synchronous operation on public data networks
X.25	CCITT standard governing interface between DTE and DCE device for terminals operating in the packet mode on public data networks
X.25 Pad	a device that permits communication between non X.25 devices and the devices in an X.25 network
X.3/X.28/X.29	a set of internationally agreed standard protocols defined to allow a character oriented device, such as a visual display terminal, to be connected to a packet switched data network
X-ON/X-OFF	*transmitter on/transmitter off* – control characters used for flow control, instructing a terminal to start transmission (X-ON or Control-S) and end transmission (X-OFF or Control-Q)

Appendix B

Interface standards

PIN NO	DB-9 CONNECTOR	DB-25 CONNECTOR	DB-25 CONNECTOR	
	EIA-232 Pin Assignment	EIA-232 Pin Assignment	EIA/TIA-530 Pin Assignment	
1	Received Line Signal	Shield	Shield	
2	Received Data	Transmitted Data	Transmitted Data	(A)
3	Transmitted Data	Received Data	Received Data	(A)
4	DTE Ready	Request to Send	Request to Send	(A)
5	Signal Common/Ground	Clear to Send	Clear to Send	(A)
6	DCE Ready	DCE Ready	DCE Ready	(A)
7	Request to Send	Signal Ground/Common	Signal Ground/Common	
8	Clear to Send	Received Line Signal	Received Line Signal	(A)
9	Ring Indicator	+ Voltage (testing)	Receiver Signal DCE Element Timing	(B)

Table B.1
Table of the common DB-9 & DB-25 pin assignments for EIA-232 and EIA/TIA-530 (often used for EIA-422 andEIA-485)
(continued over page)

10		– Voltage (testing)	Received Line	(B)
11		Unassigned	Transmitter Signal DTE Element Timing	(B)
12		Sec Received Line Signal Detector/Data Signal	Transmitter Signal DCE Element Timing	(B)
13		Sec Clear to Send	Clear to Send	(B)
14		Sec Transmitted Data	Transmitted Data	(B)
15		Transmitter Signal DCE Element Timing	Transmitter Signal DCE Element Timing	
16		Sec Received Data	Received Data	(B)
17		Receiver Signal DCE Element Timing	Receiver Signal DCE Element Timing	(A)
18		Local Loopback	Local Loopback	
19		Sec Request to Send	Request to Send	(B)
20		DTE Ready	DTE Ready	(A)
21		Remote Loopback/ Signal Quality Detector	Remote Loopback	
22		Ring Indicator	DCE Ready	(B)
23		Data Signal Rate	DTE Ready	(B)
24		Transmit Signal DTE Element Timing	Transmitter Signal DTE Element Timing	(A)
25		Test Mode	Test Mode	

Table B.1
Table of the common DB-9 & DB-25 pin assignments for EIA-232 and EIA/TIA-530 (often used for EIA-422 and EIA-485)

PIN NO	DB-9 CONNECTOR		DB-37 CONNECTOR	
	Common for EIA-422 & RIA 485		EIA-449 Pin Assignment	
1	Shield		Shield	
2	Send Data	(B+)	Signaling Rate Indic	
3	Receive Data	(B+)		
4	Request to Send	(B+)	Send Data	(A–)
5	Clear to Send	(B+)	Send Timing	(A–)
6	Send Data	(A–)	Receive Data	(A–)
7	Receive Data	(A–)	Request to Send	(A–)
8	Request to Send	(A–)	Receive Timing	(A–)
9	Clear to Send	(A–)	Clear to Send	(A–)
10			Local Loopback	
11			Data Mode	(A–)
12			Terminal Ready	(A–)
13			Receiver Ready	(A–)
14			Remote Loopback	
15			Incoming Call	
16			Select Frequency	
17			Terminal Timing	(A–)
18			Test Mode	
19			Signal Ground	
20			Receive Common	
21				
22			Send Data	(B+)
23			Send Timing	(B+)
24			Receive Data	(B+)
25			Request to Send	(B+)

Table B.2
Common DB-9 pin assignments for EIA-422 and EIA-485 and DB-37 pin assignments specified according to EIA-449
(continued over page).

26		Receive Timing	(B+)
27		Clear to Send	(B+)
28		Terminal in Service	(B+)
29		Data Mode	(B+)
30		Terminal Ready	(B+)
31		Receiver Ready	
32		Select Standby	
33		Signal Quality	
34		New Signal	
35		Terminal Timing	(B+)
36		Standby/Indicator	
37		Send Common	

Table B.2
Common DB-9 pin assignments for EIA-422 and EIA-485 and DB-37 pin assignments specified according to EIA-449

Note: EIA-449 (CCITT V.35) defines the electrical and mechanical DTE/DCE interface and was originally intended to replace EIA-232. These pin assignments are often used with EIA-422 or EIA-485 when a DB-37 connector is used.

Appendix C
CITECT practical

Objective

For Citect Version 5 Edit 3 (PROJECT NEW)

The Citect practical is designed to introduce the Citect software. This is accomplished by helping you set up, design, and run a working SCADA system. The system when running will display a main menu with seven buttons labeled TANK, ALARM, HARDWARE, TRENDS, SUMMARY, DISABLED and SHUTDOWN. You will be able to push any of the buttons and view the appropriate pages. The TANK page will show a working tank filling device. You will be able to move a tank fill switch and the tank on the screen will show a graph of the new tank level. If the tank is over-filled or under-filled an alarm will show on the screen. You will be able to view and clear the alarms in the ALARM and SUMMARY pages. You will be able to view a graph of the levels versus time in the TREND page. The SHUTDOWN button can be pressed to exit out of the system. The DISABLED page is used to disable alarms.

Citect design information

The Citect package is setup in a page format. Each page has to be opened, defined, and saved to a project. The pages used in this practical are kept simple because of time constraints. The Citect graphics builder and project editor are used to either design the pages or set up the software or compile and run the project.

The flow of the project will be as follows. (Note the software must be setup in the following order.)

Hint – If you have any problems press the F1 key on the keyboard. This will show you the help screen.

Opening a new project

Open a Citect by clicking on the Citect icon on the desktop. Click on file then new project.

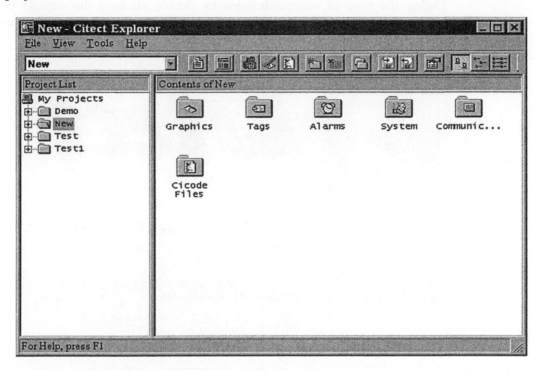

Type new. If this name is already used then put in one of your choosing. The screen should appear as follows (Add in the name new)

Then press OK.

Do not press return. The 'new' project is now created.

Defining communications

Open the project editor by clicking on tools then project editor in the Citect explorer page.

Define the type of communications by double clicking the express wizard in the communications menu. Follow the program to include the following selections;

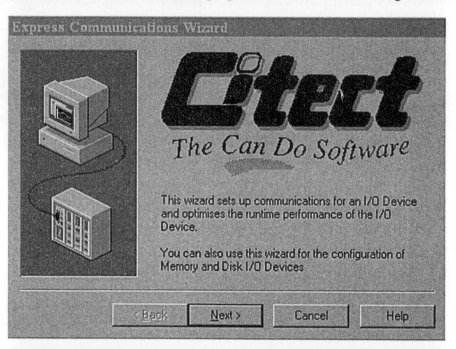

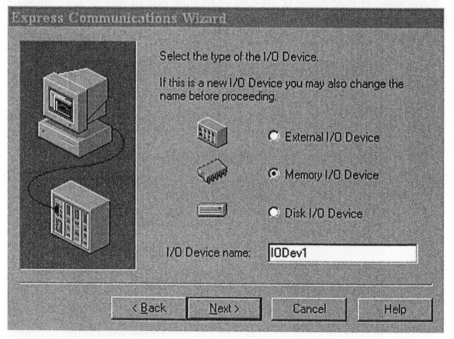

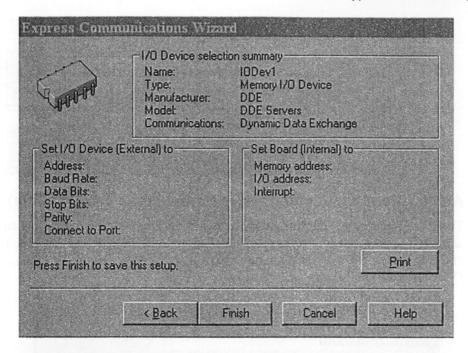

Defining tags

Go to the project editor. Click on tags on the menu. Setup your variable tag as shown and after the information is entered press add. Do not press return.

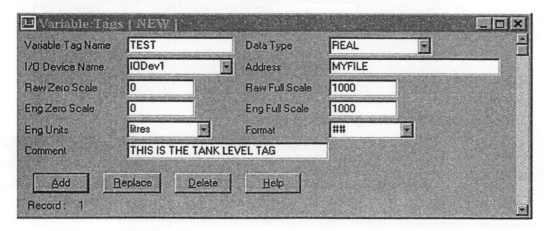

Trend tags

Still in the project editor, click on tags then trends tag on the menu. Setup your trends tag as shown and after the information is entered press add.

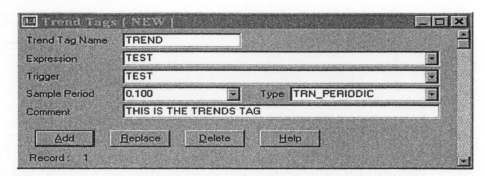

Analog alarm tags (found under alarms in the project editor)

There are four analog alarm tags. These tags define the low, high, low low and high high alarms. The low alarm is set up for 250. The high alarm is set up for 750. The low low alarm is set for 100 and the high high alarm is set for 900. Set up the four analog alarms. Do this for all four alarms.

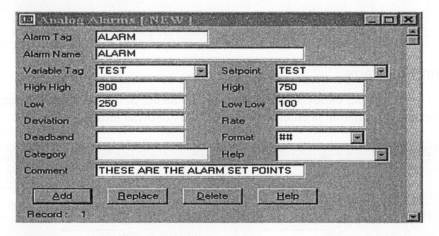

Creating the graphic pages

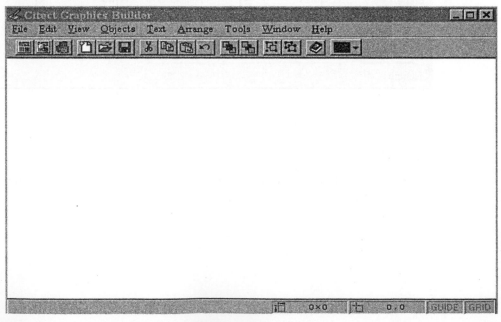

To change to the graphics builder click on tools then graphics builder. The graphic editor is used to add graphics pages to your project. You will create the pages (TANK, SUMMARY, DISABLED, ALARM, HARDWARE, TRENDS pages) in the graphics builder. Double click on new under the file menu. The screen should look like this:

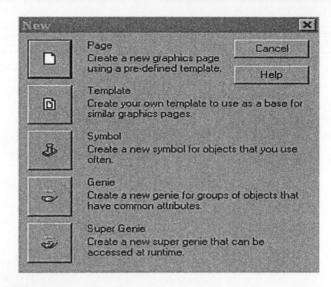

Click on Page. The screen should change to:

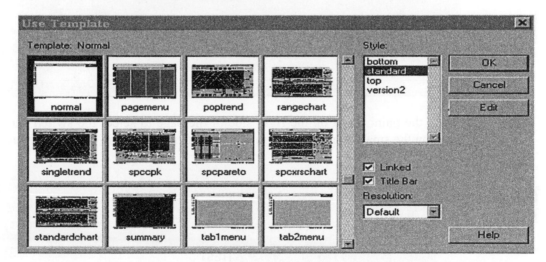

Select the following pages and save them with the names provided. After opening each page, save and name each page with the following names.

ALARM   -   ALARM
HARDWARE   -   HARDWARE
NORMAL   -   TANK
SINGLE TREND   -   TRENDS
DISABLED   -   DISABLED
SUMMARY   -   SUMMARY

The only pages that need to be modified are the tank and trends pages. When all pages have been created and saved, open the tank page.

Building the tank page

The tank page will look like this when it is done.

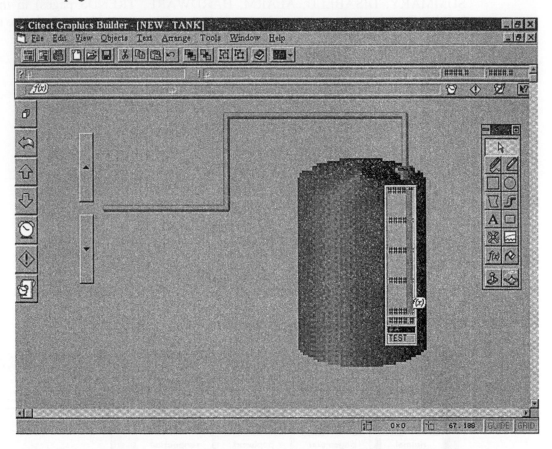

Use the paint menu on the right side of the page to select the switch and tank graph. Click on the small lamp icon and setup the switch as shown in the next picture.

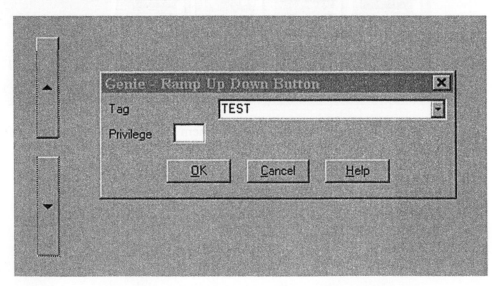

Set up the tank by clicking on the small rubber stamp icon as shown in the next picture.

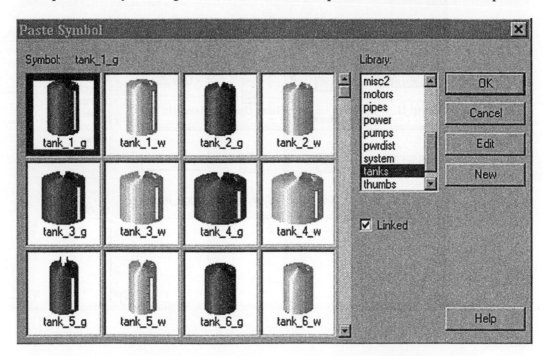

Set up the panel on the tank by selecting the panel under the lamp icon.

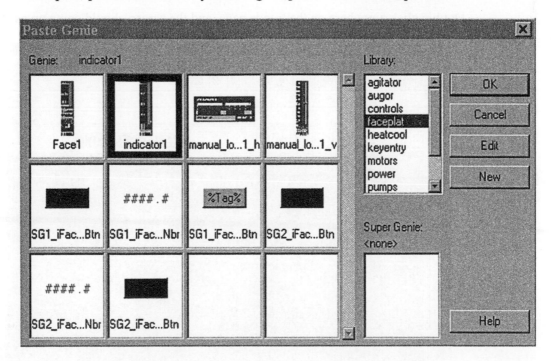

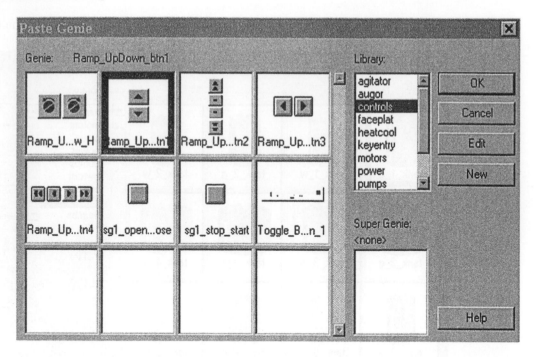

Selecting the trends page

Double click anywhere on the trends page, and the screen should come up as shown.

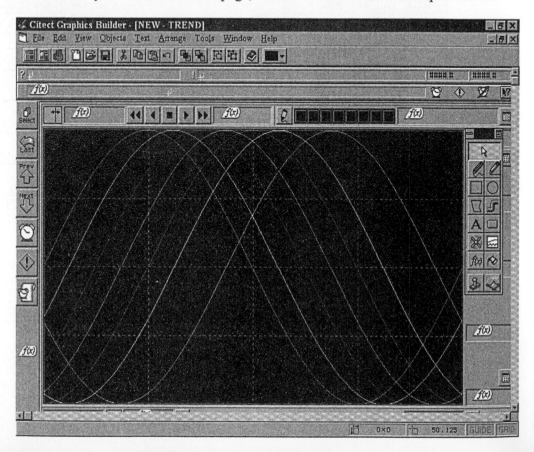

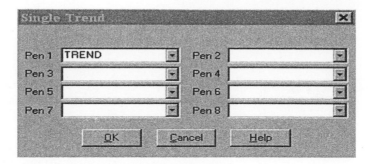

Double click again the trends page and the following page will come up. Set up as shown.

Compiling and running the project

Click on file at the top of the page and then again on compile. When the project has compiled click on file and then run.

When the program is running click on the switches and notice the tank graph will fill or empty. Notice the alarm at the top of the screen. Click on the flashing clock to view the alarm text. Click on the text and the alarm will be canceled. To view the trends select the trends page.

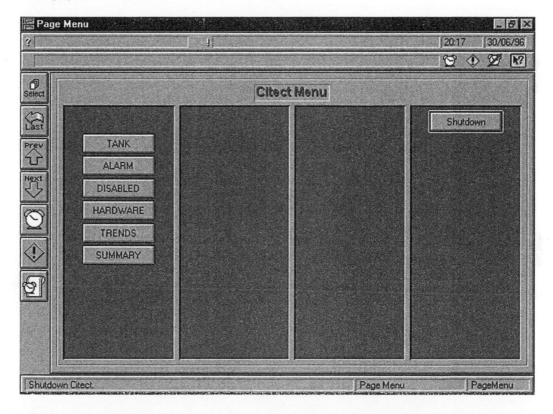

Click and hold the tank button while moving the button up and down. Notice that the graph moves and if the level is over or under limits, the alarm will show up on the top of the screen. View the alarms by clicking on the flashing alarm clock on the top right of the screen. Also, view the trends by going back to the main menu and clicking on trends.

Index

Printed and bound by CPI Group (UK) Ltd, Croydon, CR0 4YY

03/10/2024

01040338-0016